History of Astronomy

Compiled by
Shelley Gunn

Scribbles

Year of Publication 2018

ISBN : 9789387513044

Book Published by

Scribbles

(An Imprint of Alpha Editions)

email - alphaedis@gmail.com

Produced by: PediaPress GmbH
Limburg an der Lahn
Germany
http://pediapress.com/

Contents

Prehistoric

Archaeoastronomy

Archaeoastronomy (also spelled **archeoastronomy**) is the study of how people in the past "have understood the phenomena in the sky, how they used these phenomena and what role the sky played in their cultures."[1] Clive Ruggles argues it is misleading to consider archaeoastronomy to be the study of ancient astronomy, as modern astronomy is a scientific discipline, while archaeoastronomy considers symbolically rich cultural interpretations of phenomena in the sky by other cultures.[2,3] It is often twinned with *ethnoastronomy*, the anthropological study of skywatching in contemporary societies. Archaeoastronomy is also closely associated with historical astronomy, the use of historical records of heavenly events to answer astronomical problems and the history of astronomy, which uses written records to evaluate past astronomical practice.

Archaeoastronomy uses a variety of methods to uncover evidence of past practices including archaeology, anthropology, astronomy, statistics and probability, and history. Because these methods are diverse and use data from such different sources, integrating them into a coherent argument has been a long-term difficulty for archaeoastronomers.[4] Archaeoastronomy fills complementary niches in landscape archaeology and cognitive archaeology. Material evidence and its connection to the sky can reveal how a wider landscape can be integrated into beliefs about the cycles of nature, such as Mayan astronomy and its relationship with agriculture.[5] Other examples which have brought together ideas of cognition and landscape include studies of the cosmic order embedded in the roads of settlements.[6,7]

Archaeoastronomy can be applied to all cultures and all time periods. The meanings of the sky vary from culture to culture; nevertheless there are scientific methods which can be applied across cultures when examining ancient beliefs.[8] It is perhaps the need to balance the social and scientific aspects of archaeoastronomy which led Clive Ruggles to describe it as: "...[A] *field with*

Figure 1: *The rising Sun illuminates the inner chamber of Newgrange, Ireland, only at the winter solstice.*

academic work of high quality at one end but uncontrolled speculation bordering on lunacy at the other."[9]

History of archaeoastronomy

> *In his short history of 'Astro-archaeology' John Michell argued that the status of research into ancient astronomy had improved over the past two centuries, going 'from lunacy to heresy to interesting notion and finally to the gates of orthodoxy.' Nearly two decades later, we can still ask the question: Is archaeoastronomy still waiting at the gates of orthodoxy or has it gotten inside the gates?*
>
> —*Todd Bostwick quoting John Michell*[10]

Two hundred years before Michell wrote the above, there were no archaeoastronomers and there were no professional archaeologists, but there were astronomers and antiquarians. Some of their works are considered precursors of archaeoastronomy; antiquarians interpreted the astronomical orientation of the ruins that dotted the English countryside as William Stukeley did of Stonehenge in 1740,[11] while John Aubrey in 1678[12] and Henry Chauncy in 1700 sought similar astronomical principles underlying the orientation of

Figure 2: *Early archaeoastronomy surveyed Megalithic constructs in the British Isles, at sites like Auglish in County Londonderry, in an attempt to find statistical patterns*

churches.[13] Late in the nineteenth century astronomers such as Richard Proctor and Charles Piazzi Smyth investigated the astronomical orientations of the pyramids.[14]

The term *archaeoastronomy* was first used by Elizabeth Chesley Baity (at the suggestion of Euan MacKie) in 1973,[15] but as a topic of study it may be much older, depending on how archaeoastronomy is defined. Clive Ruggles[16] says that Heinrich Nissen, working in the mid-nineteenth century was arguably the first archaeoastronomer. Rolf Sinclair[17] says that Norman Lockyer, working in the late 19th and early 20th centuries, could be called the 'father of archaeoastronomy.' Euan MacKie[18] would place the origin even later, stating: "...the genesis and modern flowering of archaeoastronomy must surely lie in the work of Alexander Thom in Britain between the 1930s and the 1970s."

In the 1960s the work of the engineer Alexander Thom and that of the astronomer Gerald Hawkins, who proposed that Stonehenge was a Neolithic computer,[19] inspired new interest in the astronomical features of ancient sites. The claims of Hawkins were largely dismissed,[20] but this was not the case for Alexander Thom's work, whose survey results of megalithic sites hypothesized widespread practice of accurate astronomy in the British Isles.[21] Euan

Figure 3: *It has been proposed that Maya sites such as Uxmal were built in accordance with astronomical alignments.*

MacKie, recognizing that Thom's theories needed to be tested, excavated at the Kintraw standing stone site in Argyllshire in 1970 and 1971 to check whether the latter's prediction of an observation platform on the hill slope above the stone was correct. There was an artificial platform there and this apparent verification of Thom's long alignment hypothesis (Kintraw was diagnosed as an accurate winter solstice site) led him to check Thom's geometrical theories at the Cultoon stone circle in Islay, also with a positive result. MacKie therefore broadly accepted Thom's conclusions and published new prehistories of Britain.[22] In contrast a re-evaluation of Thom's fieldwork by Clive Ruggles argued that Thom's claims of high accuracy astronomy were not fully supported by the evidence.[23] Nevertheless, Thom's legacy remains strong, Krupp[24] wrote in 1979, "Almost singlehandedly he has established the standards for archaeo-astronomical fieldwork and interpretation, and his amazing results have stirred controversy during the last three decades." His influence endures and practice of statistical testing of data remains one of the methods of archaeoastronomy.[25,26]

The approach in the New World, where anthropologists began to consider more fully the role of astronomy in Amerindian civilizations, was markedly different. They had access to sources that the prehistory of Europe lacks such as ethnographies[27,28] and the historical records of the early colonizers. Following the pioneering example of Anthony Aveni,[29,30] this allowed New World archaeoastronomers to make claims for motives which in the Old World would have been mere speculation. The concentration on historical data led to some claims of high accuracy that were comparatively weak when compared to the statistically led investigations in Europe.[31]

This came to a head at a meeting sponsored by the International Astronomical Union (IAU) in Oxford in 1981.[32] The methodologies and research questions of the participants were considered so different that the conference proceedings were published as two volumes.[33,34] Nevertheless, the conference was considered a success in bringing researchers together and Oxford conferences have continued every four or five years at locations around the world. The subsequent conferences have resulted in a move to more interdisciplinary approaches with researchers aiming to combine the contextuality of archaeological research,[35] which broadly describes the state of archaeoastronomy today, rather than merely establishing the existence of ancient astronomies, archaeoastronomers seek to explain why people would have an interest in the night sky.

Relations to other disciplines

> *...[O]ne of the most endearing characteristics of archaeoastronomy is its capacity to set academics in different disciplines at loggerheads with each other.*
>
> —*Clive Ruggles*[36]

Archaeoastronomy has long been seen as an interdisciplinary field that uses written and unwritten evidence to study the astronomies of other cultures. As such, it can be seen as connecting other disciplinary approaches for investigating ancient astronomy: astroarchaeology (an obsolete term for studies that draw astronomical information from the alignments of ancient architecture and landscapes), history of astronomy (which deals primarily with the written textual evidence), and ethnoastronomy (which draws on the ethnohistorical record and contemporary ethnographic studies).[37,38]

Reflecting Archaeoastronomy's development as an interdisciplinary subject, research in the field is conducted by investigators trained in a wide range of disciplines. Authors of recent doctoral dissertations have described their work as concerned with the fields of archaeology and cultural anthropology; with various fields of history including the history of specific regions and periods, the history of science and the history of religion; and with the relation of astronomy to art, literature and religion. Only rarely did they describe their work as astronomical, and then only as a secondary category.[39]

Both practicing archaeoastronomers and observers of the discipline approach it from different perspectives. George Gummerman and Miranda Warburton view archaeoastronomy as part of an archaeology informed by cultural anthropology and aimed at understanding a "group's conception of themselves in relation to the heavens', in a word, its cosmology.[40] Todd Bostwick argued that "archaeoastronomy is anthropology – the study of human behavior in the past and present."[41] Paul Bahn has described archaeoastronomy as an area of cognitive archaeology.[42] Other researchers relate archaeoastronomy to the history of science, either as it relates to a culture's observations of nature and the conceptual framework they devised to impose an order on those observations[43] or as it relates to the political motives which drove particular historical actors to deploy certain astronomical concepts or techniques.[44,45] Art historian Richard Poss took a more flexible approach, maintaining that the astronomical rock art of the North American Southwest should be read employing "the hermeneutic traditions of western art history and art criticism"[46] Astronomers, however, raise different questions, seeking to provide their students with identifiable precursors of their discipline, and are especially concerned with the important question of how to confirm that specific sites are, indeed, intentionally astronomical.[47]

The reactions of professional archaeologists to archaeoastronomy have been decidedly mixed. Some expressed incomprehension or even hostility, varying from a rejection by the archaeological mainstream of what they saw as an archaeoastronomical fringe to an incomprehension between the cultural focus of archaeologists and the quantitative focus of early archaeoastronomers.[48] Yet archaeologists have increasingly come to incorporate many of the insights from archaeoastronomy into archaeology textbooks[49] and, as mentioned above, some students wrote archaeology dissertations on archaeoastronomical topics.

Since archaeoastronomers disagree so widely on the characterization of the discipline, they even dispute its name. All three major international scholarly associations relate archaeoastronomy to the study of culture, using the term *Astronomy in Culture* or a translation. Michael Hoskin sees an important part of the discipline as fact-collecting, rather than theorizing, and proposed to label this aspect of the discipline *Archaeotopography*.[50] Ruggles and Saunders proposed *Cultural Astronomy* as a unifying term for the various methods of studying folk astronomies.[51] Others have argued that astronomy is an inaccurate term, what are being studied are cosmologies and people who object to the use of logos have suggested adopting the Spanish *cosmovisión*.[52]

When debates polarise between techniques, the methods are often referred to by a colour code, based on the colours of the bindings of the two volumes from the first Oxford Conference, where the approaches were first distinguished.[53] Green (Old World) archaeoastronomers rely heavily on statistics and are sometimes accused of missing the cultural context of what is a social practice. Brown (New World) archaeoastronomers in contrast have abundant ethnographic and historical evidence and have been described as 'cavalier' on matters of measurement and statistical analysis.[54] Finding a way to integrate various approaches has been a subject of much discussion since the early 1990s.[55,56]

Methodology

> *For a long time I have believed that such diversity requires the invention of some all-embracing theory. I think I was very naïve in thinking that such a thing was ever possible.*
>
> —*Stanislaw Iwaniszewski*[57]

There is no one way to do Archaeoastronomy. The divisions between archaeoastronomers tend not to be between the physical scientists and the social scientists. Instead it tends to depend on the location of kind of data available to the researcher. In the Old World, there is little data but the sites themselves; in the New World, the sites were supplemented by ethnographic and historic data. The effects of the isolated development of archaeoastronomy in different places can still often be seen in research today. Research methods can be classified as falling into one of two approaches, though more recent projects often use techniques from both categories.

Green archaeoastronomy

Green Archaeoastronomy is named after the cover of the book *Archaeoastronomy in the Old World*.[58] It is based primarily on statistics and is particularly apt for prehistoric sites where the social evidence is relatively scant compared to the historic period. The basic methods were developed by Alexander Thom during his extensive surveys of British megalithic sites.

Thom wished to examine whether or not prehistoric peoples used high-accuracy astronomy. He believed that by using horizon astronomy, observers could make estimates of dates in the year to a specific day. The observation required finding a place where on a specific date the sun set into a notch on the horizon. A common theme is a mountain which blocked the Sun, but on the right day would allow the tiniest fraction to re-emerge on the other side for a 'double sunset'. The animation below shows two sunsets at a hypothetical site, one the day before the summer solstice and one at the summer solstice, which has a double sunset.

To test this idea he surveyed hundreds of stone rows and circles. Any individual alignment could indicate a direction by chance, but he planned to show that together the distribution of alignments was non-random, showing that there was an astronomical intent to the orientation of at least some of the alignments. His results indicated the existence of eight, sixteen, or perhaps even thirty-two approximately equal divisions of the year. The two solstices, the two equinoxes and four cross-quarter days, days half-way between a solstice and the equinox were associated with the medieval Celtic calendar.[59] While not all these conclusions have been accepted, it has had an enduring influence on archaeoastronomy, especially in Europe.

Euan MacKie has supported Thom's analysis, to which he added an archaeological context by comparing Neolithic Britain to the Mayan civilization to argue for a stratified society in this period. To test his ideas he conducted a couple of excavations at proposed prehistoric observatories in Scotland. Kintraw is a site notable for its four-meter high standing stone. Thom proposed that this was a foresight to a point on the distant horizon between Beinn Shianaidh and Beinn o'Chaolias on Jura.[60] This, Thom argued, was a notch on the horizon where a double sunset would occur at midwinter. However, from ground level, this sunset would be obscured by a ridge in the landscape, and the viewer would need to be raised by two meters: another observation platform was needed. This was identified across a gorge where a platform was formed from small stones. The lack of artifacts caused concern for some archaeologists and the petrofabric analysis was inconclusive, but further research at Maes Howe[61] and on the Bush Barrow Lozenge[62] led MacKie to conclude that while the term 'science' may be anachronistic, Thom was broadly correct upon the subject of high-accuracy alignments.[63]

In contrast Clive Ruggles has argued that there are problems with the selection of data in Thom's surveys.[64,65] Others have noted that the accuracy of horizon astronomy is limited by variations in refraction near the horizon.[66] A deeper criticism of Green archaeoastronomy is that while it can answer *whether* there was likely to be an interest in astronomy in past times, its lack of a social element means that it struggles to answer *why* people would be interested, which makes it of limited use to people asking questions about the society of the past. Keith Kintigh wrote: "To put it bluntly, in many cases it doesn't matter much to the progress of anthropology whether a particular archaeoastronomical claim is right or wrong because the information doesn't inform the current interpretive questions."[67] Nonetheless the study of alignments remains a staple of archaeoastronomical research, especially in Europe.[68]

Figure 4: *"El Caracol" a possible observatory temple at Chichen Itza.*

Brown archaeoastronomy

In contrast to the largely alignment-oriented statistically led methods of Green archaeoastronomy, Brown archaeoastronomy has been identified as being closer to the history of astronomy or to cultural history, insofar as it draws on historical and ethnographic records to enrich its understanding of early astronomies and their relations to calendars and ritual. The many records of native customs and beliefs made by the Spanish chroniclers means that Brown archaeoastronomy is most often associated with studies of astronomy in the Americas.[69]

One famous site where historical records have been used to interpret sites is Chichen Itza. Rather than analysing the site and seeing which targets appear popular, archaeoastronomers have instead examined the ethnographic records to see what features of the sky were important to the Mayans and then sought archaeological correlates. One example which could have been overlooked without historical records is the Mayan interest in the planet Venus. This interest is attested to by the Dresden codex which contains tables with information about the Venus's appearances in the sky.[70] These cycles would have been of astrological and ritual significance as Venus was associated with Quetzalcoatl or Xolotl.[71] Associations of architectural features with settings of Venus can be found in Chichen Itza, Uxmal, and probably some other Mesoamerican sites.

The Temple of the Warriors bears iconography depicting feathered serpents associated with Quetzalcoatl or Kukulcan. This means that the building's alignment towards the place on the horizon where Venus first appears in the evening sky (when it coincides with the rainy season) may be meaningful.[72] However, since both the date and the azimuth of this event change continuously, a solar interpretation of this orientation is much more likely.

Aveni claims that another building associated with the planet Venus in the form of Kukulcan, and the rainy season at Chichen Itza is the Caracol.[73] This is a building with circular tower and doors facing the cardinal directions. The base faces the most northerly setting of Venus. Additionally the pillars of a stylobate on the building's upper platform were painted black and red. These are colours associated with Venus as an evening and morning star.[74] However the windows in the tower seem to have been little more than slots, making them poor at letting light in, but providing a suitable place to view out.[75]

Aveni states that one of the strengths of the Brown methodology is that it can explore astronomies invisible to statistical analysis and offers the astronomy of the Incas as another example. The empire of the Incas was conceptually divided using *ceques* radial routes emanating from the capital at Cusco. Thus there are alignments in all directions which would suggest there is little of astronomical significance, However, ethnohistorical records show that the various directions do have cosmological and astronomical significance with various points in the landscape being significant at different times of the year.[76,77] In eastern Asia archaeoastronomy has developed from the History of Astronomy and much archaeoastronomy is searching for material correlates of the historical record. This is due to the rich historical record of astronomical phenomena which, in China, stretches back into the Han dynasty, in the second century BC.[78]

A criticism of this method is that it can be statistically weak. Schaefer in particular has questioned how robust the claimed alignments in the Caracol are.[79,80] Because of the wide variety of evidence, which can include artefacts as well as sites, there is no one way to practice archaeoastronomy.[81] Despite this it is accepted that archaeoastronomy is not a discipline that sits in isolation. Because archaeoastronomy is an interdisciplinary field, whatever is being investigated should make sense both archaeologically and astronomically. Studies are more likely to be considered sound if they use theoretical tools found in archaeology like analogy and homology and if they can demonstrate an understanding of accuracy and precision found in astronomy. Both quantitative analyses and interpretations based on ethnographic analogies and other contextual evidence have recently been applied in systematic studies of architectural orientations in the Maya area and in other parts of Mesoamerica.

Source materials

Because archaeoastronomy is about the many and various ways people interacted with the sky, there are a diverse range of sources giving information about astronomical practices.

Alignments

A common source of data for archaeoastronomy is the study of alignments. This is based on the assumption that the axis of alignment of an archaeological site is meaningfully oriented towards an astronomical target. Brown archaeoastronomers may justify this assumption through reading historical or ethnographic sources, while Green archaeoastronomers tend to prove that alignments are unlikely to be selected by chance, usually by demonstrating common patterns of alignment at multiple sites.

An alignment is calculated by measuring the azimuth, the angle from north, of the structure and the altitude of the horizon it faces[82] The azimuth is usually measured using a theodolite or a compass. A compass is easier to use, though the deviation of the Earth's magnetic field from true north, known as its magnetic declination must be taken into account. Compasses are also unreliable in areas prone to magnetic interference, such as sites being supported by scaffolding. Additionally a compass can only measure the azimuth to a precision of a half a degree.

A theodolite can be considerably more accurate if used correctly, but it is also considerably more difficult to use correctly. There is no inherent way to align a theodolite with North and so the scale has to be calibrated using astronomical observation, usually the position of the Sun.[83] Because the position of celestial bodies changes with the time of day due to the Earth's rotation, the time of these calibration observations must be accurately known, or else there will be a systematic error in the measurements. Horizon altitudes can be measured with a theodolite or a clinometer.

Artifacts

For artifacts such as the Sky Disc of Nebra, alleged to be a Bronze Age artefact depicting the cosmos,[84,85] the analysis would be similar to typical post-excavation analysis as used in other sub-disciplines in archaeology. An artefact is examined and attempts are made to draw analogies with historical or ethnographical records of other peoples. The more parallels that can be found, the more likely an explanation is to be accepted by other archaeologists.

A more mundane example is the presence of astrological symbols found on some shoes and sandals from the Roman Empire. The use of shoes and sandals

Figure 5: *The Antikythera mechanism (main fragment)*

is well known, but Carol van Driel-Murray has proposed that astrological symbols etched onto sandals gave the footwear spiritual or medicinal meanings.[86] This is supported through citation of other known uses of astrological symbols and their connection to medical practice and with the historical records of the time.Wikipedia:Citation needed

Another well-known artefact with an astronomical use is the Antikythera mechanism. In this case analysis of the artefact, and reference to the description of similar devices described by Cicero, would indicate a plausible use for the device. The argument is bolstered by the presence of symbols on the mechanism, allowing the disc to be read.[87]

Art and inscriptions

Art and inscriptions may not be confined to artefacts, but also appear painted or inscribed on an archaeological site. Sometimes inscriptions are helpful enough to give instructions to a site's use. For example, a Greek inscription on a stele (from Itanos) has been translated as:"Patron set this up for Zeus Epopsios. Winter solstice. Should anyone wish to know: off 'the little pig' and the stele the sun turns."[88] From Mesoamerica come Mayan and Aztec codices. These are folding books made from Amatl, processed tree bark on which are glyphs

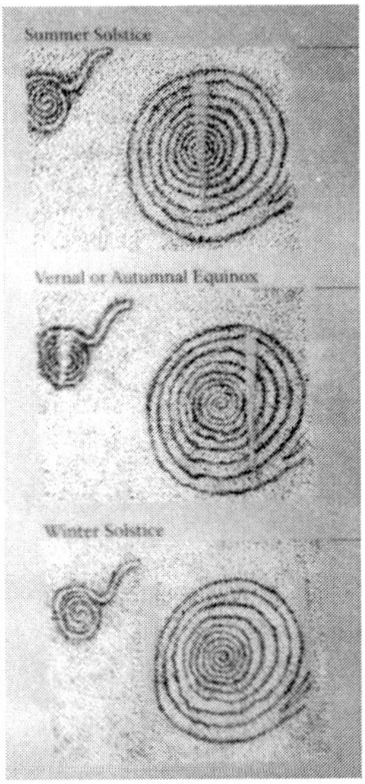

Figure 6: *Diagram showing the location of the sun daggers on the Fajada Butte petroglyph on various days*

in Mayan or Aztec script. The Dresden codex contains information regarding the Venus cycle, confirming its importance to the Mayans.

More problematic are those cases where the movement of the Sun at different times and seasons causes light and shadow interactions with petroglyphs.[89] A widely known example is the Sun Dagger of Fajada Butte at which a glint of sunlight passes over a spiral petroglyph.[90] The location of a dagger of light on the petroglyph varies throughout the year. At the summer solstice a dagger can be seen through the heart of the spiral; at the winter solstice two daggers appear to either side of it. It is proposed that this petroglyph was created to mark these events. Recent studies have identified many similar sites in the US Southwest and Northwestern Mexico.[91,92] It has been argued that the number of solstitial markers at these sites provides statistical evidence that they were intended to mark the solstices.[93] The Sun Dagger site on Fajada Butte in Chaco Canyon, New Mexico, stands out for its explicit light markings that record all the key events of both the solar and lunar cycles: summer solstice, winter solstice, equinox, and the major and minor lunar standstills of the moon's 18.6 year cycle.[94,95] In addition at two other sites on Fajada Butte, there are five light markings on petroglyphs recording the summer and winter solstices, equinox and solar noon.[96] Numerous buildings and interbuilding alignments of the great houses of Chaco Canyon and outlying areas are oriented to the same solar and lunar directions that are marked at the Sun Dagger site.[97]

If no ethnographic nor historical data are found which can support this assertion then acceptance of the idea relies upon whether or not there are enough petroglyph sites in North America that such a correlation could occur by chance. It is helpful when petroglyphs are associated with existing peoples. This allows ethnoastronomers to question informants as to the meaning of such symbols.

Ethnographies

As well as the materials left by peoples themselves, there are also the reports of other who have encountered them. The historical records of the Conquistadores are a rich source of information about the pre-Columbian Americans. Ethnographers also provide material about many other peoples.

Aveni uses the importance of zenith passages as an example of the importance of ethnography. For peoples living between the tropics of Cancer and Capricorn there are two days of the year when the noon Sun passes directly overhead and casts no shadow. In parts of Mesoamerica this was considered a significant day as it would herald the arrival of rains, and so play a part in the cycle of agriculture. This knowledge is still considered important amongst Mayan Indians living in Central America today. The ethnographic records suggested to archaeoastronomers that this day may have been important to the ancient Mayans. There are also shafts known as 'zenith tubes' which illuminate subterranean rooms when the sun passes overhead found at places like Monte Albán and Xochicalco. It is only through the ethnography that we can speculate that the timing of the illumination was considered important in Mayan society.[98] Alignments to the sunrise and sunset on the day of the zenith passage have been claimed to exist at several sites. However, it has been shown that, since there are very few orientations that can be related to these phenomena, they likely have different explanations.[99]

Ethnographies also caution against over-interpretation of sites. At a site in Chaco Canyon can be found a pictograph with a star, crescent and hand. It has been argued by some astronomers that this is a record of the 1054 Supernova.[100] However recent reexaminations of related 'supernova petroglyphs' raises questions about such sites in general[101] and anthropological evidence suggests other inrepretations. The Zuni people, who claim a strong ancestral affiliation with Chaco, marked their sun-watching station with a crescent, star, hand and sundisc, similar to those found at the Chaco site.[102]

Ethnoastronomy is also an important field outside of the Americas. For example, anthropological work with Aboriginal Australians is producing much

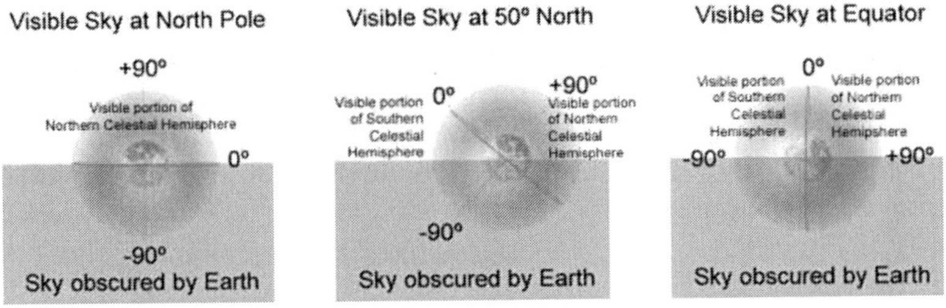

Figure 7: *Diagram of the visible portions of sky at varying latitudes.*

information about their Indigenous astronomies[103,104] and about their interaction with the modern world.[105]

Recreating the ancient sky

> ...*[A]lthough different ways to do science and different scientific results do arise in different cultures, this provides little support for those who would use such differences to question the sciences' ability to provide reliable statements about the world in which we live.*
>
> —*Stephen McCluskey*[106]

Once the researcher has data to test, it is often necessary to attempt to recreate ancient sky conditions to place the data in its historical environment.

Declination

To calculate what astronomical features a structure faced a coordinate system is needed. The stars provide such a system. If you were to go outside on a clear night you would observe the stars spinning around the celestial pole. This point is +90° if you are watching the North Celestial Pole or –90° if you are observing the Southern Celestial Pole.[107] The concentric circles the stars trace out are lines of celestial latitude, known as *declination*. The arc connecting the points on the horizon due East and due West (if the horizon is flat) and all points midway between the Celestial Poles is the Celestial Equator which has a declination of 0°. The visible declinations vary depending where you are on the globe. Only an observer on the North Pole of Earth would be unable to see any stars from the Southern Celestial Hemisphere at night (see diagram below). Once a declination has been found for the point on the horizon that a building faces it is then possible to say whether a specific body can be seen in that direction.

Solar positioning

While the stars are fixed to their declinations the Sun is not. The rising point of the Sun varies throughout the year. It swings between two limits marked by the solstices a bit like a pendulum, slowing as it reaches the extremes, but passing rapidly through the midpoint. If an archaeoastronomer can calculate from the azimuth and horizon height that a site was built to view a declination of +23.5° then he or she need not wait until 21 June to confirm the site does indeed face the summer solstice.[108] For more information see History of solar observation.

Lunar positioning

The Moon's appearance is considerably more complex. Its motion, like the Sun, is between two limits — known as *luni*stices rather than *sol*stices. However, its travel between lunistices is considerably faster. It takes a sidereal month to complete its cycle rather than the year-long trek of the Sun. This is further complicated as the lunistices marking the limits of the Moon's movement move on an 18.6 year cycle. For slightly over nine years the extreme limits of the moon are outside the range of sunrise. For the remaining half of the cycle the Moon never exceeds the limits of the range of sunrise. However, much lunar observation was concerned with the *phase* of the Moon. The cycle from one New Moon to the next runs on an entirely different cycle, the Synodic month.[109] Thus when examining sites for lunar significance the data can appear sparse due the extremely variable nature of the moon. See Moon for more details.

Stellar positioning

Finally there is often a need to correct for the apparent movement of the stars. On the timescale of human civilisation the stars have largely maintained the same position relative to each other. Each night they appear to rotate around the celestial poles due to the Earth's rotation about its axis. However, the Earth spins rather like a spinning top. Not only does the Earth rotate, it wobbles. The Earth's axis takes around 25,800 years to complete one full wobble.[110] The effect to the archaeoastronomer is that stars did not rise over the horizon in the past in the same places as they do today. Nor did the stars rotate around Polaris as they do now. In the case of the Egyptian pyramids, it has been shown they were aligned towards Thuban, a faint star in the constellation of Draco.[111] The effect can be substantial over relatively short lengths of time, historically speaking. For instance a person born on 25 December in Roman times would have been born with the sun in the constellation Capricorn. In the modern period a person born on the same date would have the sun in Sagittarius due to the precession of the equinoxes.

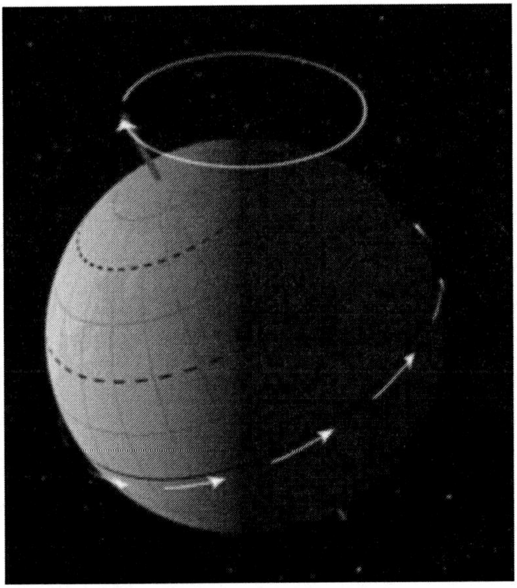

Figure 8: *Precessional movement.*

Transient phenomena

Additionally there are often transient phenomena, events which do not happen on an annual cycle. Most predictable are events like eclipses. In the case of solar eclipses these can be used to date events in the past. A solar eclipse mentioned by Herodotus enables us to date a battle between the Medes and the Lydians, which following the eclipse failed to happen, to 28 May, 585 BC. Other easily calculated events are supernovae whose remains are visible to astronomers and therefore their positions and magnitude can be accurately calculated.

Some comets are predictable, most famously Halley's Comet. Yet as a class of object they remain unpredictable and can appear at any time. Some have extremely lengthy orbital periods which means their past appearances and returns cannot be predicted. Others may have only ever passed through the Solar System once and so are inherently unpredictable.[112]

Meteor showers should be predictable, but some meteors are cometary debris and so require calculations of orbits which are currently impossible to complete.[113] Other events noted by ancients include aurorae, sun dogs and rainbows all of which are as impossible to predict as the ancient weather, but nevertheless may have been considered important phenomena.

Figure 9: *Halley's Comet depicted on the Bayeux tapestry*

Major topics of archaeoastronomical research

> *What has astronomy brought into the lives of cultural groups throughout history? The answers are many and varied...*
>
> *—Von Del Chamberlain and M. Jane Young*[114]

The use of calendars

A common justification for the need for astronomy is the need to develop an accurate calendar for agricultural reasons. Ancient texts like Hesiod's Works and Days, an ancient farming manual, would appear to contradict this. Instead astronomical observations are used in combination with ecological signs, such as bird migrations to determine the seasons. Ethnoastronomical work with the Mursi of Ethiopia shows that haphazard astronomy continued until recent times in some parts of the world.[115] All the same, calendars appear to be an almost universal phenomenon in societies as they provide tools for the regulation of communal activities.

An example of a non-agricultural calendar is the *Tzolk'in* calendar of the Maya civilization of pre-Columbian Mesoamerica, which is a cycle of 260 days. This count is based on an earlier calendar and is found throughout Mesoamerica.

Figure 10: *The constellation Argo Navis drawn by Johannes Hevelius in 1690.*

This formed part of a more comprehensive system of Maya calendars which combined a series of astronomical observations and ritual cycles.[116]

Other peculiar calendars include ancient Greek calendars. These were nominally lunar, starting with the New Moon. In reality the calendar could pause or skip days with confused citizens inscribing dates by both the civic calendar and *ton theoi*, by the moon.[117] The lack of any universal calendar for ancient Greece suggests that coordination of panhellenic events such as games or rituals could be difficult and that astronomical symbolism may have been used as a politically neutral form of timekeeping.[118] Orientation measurements in Greek temples and Byzantine churches have been associated to deity's name day, festivities, and special events.[119,120]

Myth and cosmology

Another motive for studying the sky is to understand and explain the universe. In these cultures myth was a tool for achieving this and the explanations, while not reflecting the standards of modern science, are cosmologies.

The Incas arranged their empire to demonstrate their cosmology. The capital, Cusco, was at the centre of the empire and connected to it by means of ceques, conceptually straight lines radiating out from the centre.[121] These ceques connected the centre of the empire to the four *suyus*, which were regions defined by their direction from Cusco. The notion of a quartered cosmos is common across the Andes. Gary Urton, who has conducted fieldwork in the Andean villagers of Misminay, has connected this quartering with the appearance of the Milky Way in the night sky.[122] In one season it will bisect the sky and in another bisect it in a perpendicular fashion.

The importance of observing cosmological factors is also seen on the other side of the world. The Forbidden City in Beijing is laid out to follow cosmic order though rather than observing four directions. The Chinese system was composed of five directions: North, South, East, West and Centre. The Forbidden City occupied the centre of ancient Beijing.[123] One approaches the Emperor from the south, thus placing him in front of the circumpolar stars. This creates the situation of the heavens revolving around the person of the Emperor. The Chinese cosmology is now better known through its export as feng shui.

There is also much information about how the universe was thought to work stored in the mythology of the constellations. The Barasana of the Amazon plan part of their annual cycle based on observation of the stars. When their constellation of the Caterpillar-Jaguar (roughly equivalent to the modern Scorpius) falls they prepare to catch the pupating caterpillars of the forest as they fall from the trees.[124] The caterpillars provide food at a season when other foods are scarce.[125]

A more well-known source of constellation myth are the texts of the Greeks and Romans. The origin of their constellations remains a matter of vigorous and occasionally fractious debate.[126,127]

The loss of one of the sisters, Merope, in some Greek myths may reflect an astronomical event wherein one of the stars in the Pleiades disappeared from view by the naked eye.[128]

Giorgio de Santillana, professor of the History of Science in the School of Humanities at the Massachusetts Institute of Technology, along with Hertha von Dechend believed that the old mythological stories handed down from antiquity were not random fictitious tales but were accurate depictions of celestial cosmology clothed in tales to aid their oral transmission. The chaos, monsters and violence in ancient myths are representative of the forces that shape each age. They believed that ancient myths are the remains of preliterate astronomy that became lost with the rise of the Greco-Roman civilization. Santillana and von Dechend in their book *Hamlet's Mill, An Essay on Myth and the Frame of Time* (1969) clearly state that ancient myths have no historical or factual basis other than a cosmological one encoding astronomical phenomena, especially the precession of the equinoxes.[129] Santillana and von Dechend's approach is not widely accepted.

Figure 11: *The Precinct of Amun-Re was aligned on the midwinter solstice.*

Displays of power

By including celestial motifs in clothing it becomes possible for the wearer to make claims the power on Earth is drawn from above. It has been said that the Shield of Achilles described by Homer is also a catalogue of constellations.[130] In North America shields depicted in Comanche petroglyphs appear to include Venus symbolism.[131]

Solsticial alignments also can be seen as displays of power. When viewed from a ceremonial plaza on the Island of the Sun (the mythical origin place of the Sun) in Lake Titicaca, the Sun was seen to rise at the June solstice between two towers on a nearby ridge. The sacred part of the island was separated from the remainder of it by a stone wall and ethnographic records indicate that access to the sacred space was restricted to members of the Inca ruling elite. Ordinary pilgrims stood on a platform outside the ceremonial area to see the solstice Sun rise between the towers.[132]

In Egypt the temple of Amun-Re at Karnak has been the subject of much study. Evaluation of the site, taking into account the change over time of the obliquity of the ecliptic show that the Great Temple was aligned on the rising of the midwinter sun.[133] The length of the corridor down which sunlight would travel would have limited illumination at other times of the year.

Figure 12: *The sunlight enters the tomb at Newgrange via the roofbox built above the door.*

In a later period the Serapeum in Alexandria was also said to have contained a solar alignment so that, on a specific sunrise, a shaft of light would pass across the lips of the statue of Serapis thus symbolising the Sun saluting the god.[134]

Major sites of archaeoastronomical interest

Clive Ruggles and Michel Cotte recently edited a book on heritage sites of astronomy and archaeoastronomy that provides a list of the main sites around the world.[135]

> *At Stonehenge in England and at Carnac in France, in Egypt and Yucatán, across the whole face of the earth, are found mysterious ruins of ancient monuments, monuments with astronomical significance... They mark the same kind of commitment that transported us to the moon and our spacecraft to the surface of Mars.*
>
> —*Edwin Krupp*[136]

Figure 13: *The pyramids of Giza*

Newgrange

Newgrange is a passage tomb in the Republic of Ireland dating from around 3,300 to 2,900 BC[137] For a few days around the Winter Solstice light shines along the central passageway into the heart of the tomb. What makes this notable is not that light shines in the passageway, but that it does not do so through the main entrance. Instead it enters via a hollow box above the main doorway discovered by Michael O'Kelly.[138] It is this roofbox which strongly indicates that the tomb was built with an astronomical aspect in mind. Clive Ruggles notes:

> ...[F]*ew people - archaeologists or astronomers- have doubted that a powerful astronomical symbolism was deliberately incorporated into the monument, demonstrating that a connection between astronomy and funerary ritual, at the very least, merits further investigation.*

Egypt

Since the first modern measurements of the precise cardinal orientations of the pyramids by Flinders Petrie, various astronomical methods have been proposed for the original establishment of these orientations.[139,140,141] It was recently proposed that this was done by observing the positions of two stars in the Plough / Big Dipper which was known to Egyptians as the thigh. It is thought that a vertical alignment between these two stars checked with a plumb bob

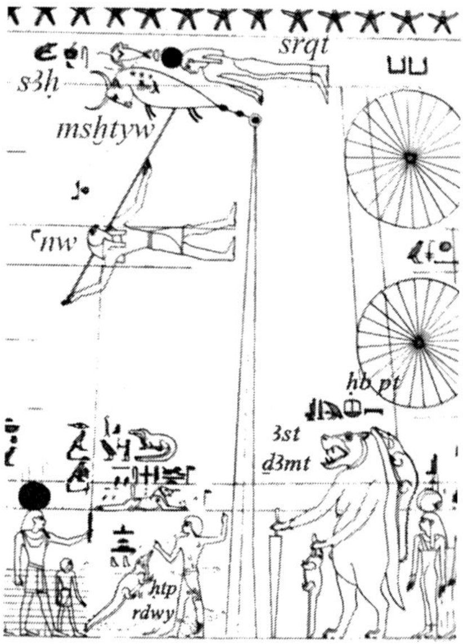

Figure 14: *Constellations on the astronomical ceiling of Senemut Tomb*

was used to ascertain where north lay. The deviations from true north using this model reflect the accepted dates of construction.[142]

Some have argued that the pyramids were laid out as a map of the three stars in the belt of Orion,[143] although this theory has been criticized by reputable astronomers.[144,145] The site was instead probably governed by a spectacular hierophany which occurs at the summer solstice, when the sun, viewed from the Sphinx terrace, forms - together with the two giant pyramids - the symbol Akhet, which was also the name of the Great Pyramid. Further, the south east corners of all the 3 pyramids align towards the temple of Heliopolis, as first discovered by the Egyptologist Mark Lehner.

The astronomical ceiling of the tomb of Senenmut (c. 1470 BC) contains the Celestial Diagram depicting circumpolar constellations in the form of discs. Each disc is divided into 24 sections suggesting a 24-hour time period. Constellations are portrayed as sacred deities of Egypt. The observation of lunar cycles is also evident.

Figure 15: *Plumed Serpent*

El Castillo

El Castillo, also known as Kukulcán's Pyramid, is a Mesoamerican step-pyramid built in the centre of Mayan center of Chichen Itza in Mexico. Several architectural features have suggested astronomical elements. Each of the stairways built into the sides of the pyramid has 91 steps. Along with the extra one for the platform at the top, this totals 365 steps, which is possibly one for each day of the year (365.25) or the number of lunar orbits in 10,000 rotations (365.01).

A visually striking effect is seen every March and September as an unusual shadow occurs around the equinoxes. A shadow appears to descend the west balustrade of the northern stairway. The visual effect is of a serpent descending the stairway, with its head at the base in light. Additionally the western face points to sunset around 25 May, traditionally the date of transition from the dry to the rainy season.[146] The intended alignment was, however, likely incorporated in the northern (main) facade of the temple, as it corresponds to sunsets on May 20 and July 24, recorded also by the central axis of Castillo at Tulum.[147]. The two dates are separated by 65 and 300 days, and it has been shown that the solar orientations in Mesoamerica regularly correspond to dates separated by calendrically significant intervals (multiples of 13 and 20 days).

Figure 16: *The sun rising over Stonehenge at the 2005 Summer Solstice.*

Stonehenge

Many astronomical alignments have been claimed for Stonehenge, a complex of megaliths and earthworks in the Salisbury Plain of England. The most famous of these is the midsummer alignment, where the Sun rises over the Heel Stone. However, this interpretation has been challenged by some archaeologists who argue that the midwinter alignment, where the viewer is outside Stonehenge and sees the sun setting in the henge, is the more significant alignment, and the midsummer alignment may be a coincidence due to local topography.[148]

As well as solar alignments, there are proposed lunar alignments. The four station stones mark out a rectangle. The short sides point towards the midsummer sunrise and midwinter sunset. The long sides if viewed towards the south-east, face the most southerly rising of the moon. Aveni notes that these lunar alignments have never gained the acceptance that the solar alignments have received.[149] The Heel Stone azimuth is one-seventh of circumference, matching the latitude of Avebury, while summer solstice sunrise azimuth is no longer equal to the construction era direction.Wikipedia:Citation needed

Figure 17: *The interior of Maeshowe chambered tomb.*

Maeshowe

This is an architecturally outstanding Neolithic chambered tomb on the Mainland of Orkney, Scotland – probably dating to the early 3rd millennium BC, and where the setting sun at midwinter shines down the entrance passage into the central chamber (see Newgrange). In the 1990s further investigations were carried out to discover whether this was an accurate or an approximate solar alignment. Several new aspects of the site were discovered. In the first place the entrance passage faces the hills of the island Hoy, about 10 miles away. Secondly, it consists of two straight lengths, angled at a few degrees to each other. Thirdly, the outer part is aligned towards the midwinter sunset position on a level horizon just to the left of Ward Hill on Hoy. Fourthly the inner part points directly at the Barnhouse standing stone about 400m away and then to the right end of the summit of Ward Hill, just before it dips down to the notch between it at Cuilags to the right. This indicated line points to sunset on the first Sixteenths of the solar year (according to A. Thom) before and after the winter solstice and the notch at the base of the right slope of the Hill is at the same declination. Fourthly a similar 'double sunset' phenomenon is seen at the right end of Cuilags, also on Hoy; here the date is the first Eighth of the year before and after the winter solstice, at the beginning of November and February respectively – the Old Celtic festivals of Samhain and Imbolc. This alignment is not indicated by an artificial structure but gains plausibility from

Figure 18: *The Palace of the Governor at Uxmal.*

the other two indicated lines. Maeshowe is thus an extremely sophisticated calendar site which must have been positioned carefully in order to use the horizon foresights in the ways described.

Uxmal

Uxmal is a Mayan city in the Puuc Hills of Yucatán Peninsula, Mexico. The Governor's Palace at Uxmal is often used as an exemplar of why it is important to combine ethnographic and alignment data. The palace is aligned with an azimuth of 118° on the pyramid of Cehtzuc. This alignment corresponds approximately to the southernmost rising and, with a much greater precision, to the northernmost setting of Venus; both phenomena occur once every eight years. By itself this would not be sufficient to argue for a meaningful connection between the two events. The palace has to be aligned in one direction or another and why should the rising of Venus be any more important than the rising of the Sun, Moon, other planets, Sirius *et cetera*? The answer given is that not only does the palace point towards significant points of Venus, it is also covered in glyphs which stand for Venus and Mayan zodiacal constellations.[150] Moreover, the great northerly extremes of Venus always occur in late April or early May, coinciding with the onset of the rainy season. The Venus glyphs placed in the cheeks of the Maya rain god Chac, most likely referring to

Figure 19: *The Great Kiva at Chaco Canyon.*

the concomitance of these phenomena, support the west-working orientation scheme.[151]

Chaco Canyon

In Chaco Canyon, the center of the ancient Pueblo culture in the American Southwest, numerous solar and lunar light markings and architectural and road alignments have been documented. These findings date to the 1977 discovery of the Sun Dagger site by Anna Sofaer.[152] Three large stone slabs leaning against a cliff channel light and shadow markings onto two spiral petroglyphs on the cliff wall, marking the solstices, equinoxes and the lunar standstills of the 18.6 year cycle of the moon. Subsequent research by the Solstice Project and others demonstrated that numerous building and interbuilding alignments of the great houses of Chaco Canyon are oriented to solar, lunar and cardinal directions.[153,154] In addition, research shows that the Great North Road, a thirty-five mile engineered "road", was constructed not for utilitarian purposes but rather to connect the ceremonial center of Chaco Canyon with the direction north.[155]

Figure 20: *According to Rappenglueck, the eyes of the bull, the bird, and the bird-man may represent the three stars Vega, Altair, and Deneb commonly known as the Summer Triangle.*

Lascaux Cave

In recent years, new research has suggested that the Lascaux cave paintings in France may incorporate prehistoric star charts. Michael Rappenglueck of the University of Munich argues that some of the non-figurative dot clusters and dots within some of the figurative images correlate with the constellations of Taurus, the Pleiades and the grouping known as the "Summer Triangle". Based on her own study of the astronomical significance of Bronze Age petroglyphs in the Vallée des Merveilles and her extensive survey of other prehistoric cave painting sites in the region—most of which appear to have been selected because the interiors are illuminated by the setting sun on the day of the winter solstice—French researcher Chantal Jègues-Wolkiewiez has further proposed that the gallery of figurative images in the Great Hall represents an extensive star map and that key points on major figures in the group correspond to stars in the main constellations as they appeared in the Paleolithic. Appliying phylogenetics to myths of the Cosmic Hunt, Julien d'Huy suggested that the palaeolithic version of this story could be the following: there is an animal that is a horned herbivore, especially an elk. One human pursues this ungulate. The hunt locates or gets to the sky. The animal is alive when it is transformed into a constellation. It forms the Big Dipper. This story may be represented in the famous Lascaux shaft 'scene'[156]

Fringe archaeoastronomy

At least now we have all the archaeological facts to go along with the astronomers, the Druids, the Flat Earthers and all the rest.

—Sir Jocelyn Stephens[157]

Archaeoastronomy owes something of this poor reputation among scholars to its occasional misuse to advance a range of pseudo-historical accounts. During the 1930s, Otto S. Reuter compiled a study entitled *Germanische Himmelskunde*, or "Teutonic Skylore". The astronomical orientations of ancient monuments claimed by Reuter and his followers would place the ancient Germanic peoples ahead of the Ancient Near East in the field of astronomy, demonstrating the intellectual superiority of the "Aryans" (Indo-Europeans) over the Semites.[158]

Since the 19th century, numerous scholars have sought to use archaeoastronomical calculations to demonstrate the antiquity of Ancient Indian Vedic culture, computing the dates of astronomical observations ambiguously described in ancient poetry to as early as 4000 BCE.[159] David Pingree, a historian of Indian astronomy, condemned "the scholars who perpetrate wild theories of prehistoric science and call themselves archaeoastronomers."[160]

More recently Gallagher,[161] Pyle,[162] and Fell[163] interpreted inscriptions in West Virginia as a description in Celtic Ogham alphabet of the supposed winter solstitial marker at the site. The controversial translation was supposedly validated by a problematic archaeoastronomical indication in which the winter solstice sun shone on an inscription of the sun at the site. Subsequent analyses criticized its cultural inappropriateness, as well as its linguistic and archeaoastronomical[164] claims, to describe it as an example of "cult archaeology".[165]

Archaeoastronomy is sometimes related to the fringe discipline of Archaeocryptography, when its followers attempt to find underlying mathematical orders beneath the proportions, size, and placement of archaeoastronomical sites such as Stonehenge and the Pyramid of Kukulcán at Chichen Itza.

Archaeoastronomical organisations and publications

There are currently three academic organisations for scholars of archaeoastronomy. ISAAC—the International Society for Archaeoastronomy and Astronomy in Culture[166]—was founded in 1995 and now sponsors the Oxford conferences and *Archaeoastronomy — the Journal of Astronomy in Culture*. SEAC— La Société Européenne pour l'Astronomie dans la Culture[167]—is slightly older; it was created in 1992. SEAC holds annual conferences in Europe and publishes refereed conference proceedings on an annual basis. There is also SIAC— La Sociedad Interamericana de Astronomía en la Cultura[168], primarily a Latin American organisation which was founded in 2003. Two new organisations focused on regional archaeoastronomy were founded in 2013: ASIA - the Australian Society for Indigenous Astronomy[169] in Australia and SMART - the Society of Māori Astronomy Research and Traditions[170] in New Zealand.

Additionally the *Journal for the History of Astronomy* publishes many archaeoastronomical papers. For twenty-seven volumes (from 1979 to 2002) it published an annual supplement *Archaeoastronomy*. The *Journal of Astronomical History and Heritage* (National Astronomical Research Institute of Thailand), *Culture & Cosmos* (University of Wales, UK) and *Mediterranean Archaeology and Archaeometry* (University of Aegean, Greece) also publish papers on archaeoastronomy.

Various national archaeoastronomical projects have been undertaken. Among them is the program at the Tata Institute of Fundamental Research named " Archaeo Astronomy in Indian Context[171]" that has made interesting findings in this field.[172]

References

Bibliography

- Aldana, G. (2007). *The Apotheosis of Janaab' Pakal: Science, History, and Religion at Classic Maya Palenque*. Boulder: University Press of Colorado. ISBN 0-87081-866-X.
- Atkinson, R.J.C. (1966). "Moonshine on Stonehenge". *Antiquity*. **49** (159): 212–6. doi: 10.1017/S0003598X0003252X[173].
- Aveni, A.F. (1979). "Astronomy in Ancient Mesoamerica". In E.C. Krupp. *In Search of Ancient Astronomies*. Chatto and Windus. pp. 154–185. ISBN 0-7011-2314-1.

- Aveni, A.F. (1980). *Skywatchers of Ancient Mexico*. University of Texas. ISBN 0-292-77578-4.
- Aveni, A.F. (1981). "Archaeoastronomy". In Michael B. Schiffer. *Advances in Archaeological Method and Theory*. **4**. Academic Press. p. 177. ISBN 0-12-003104-3.
- Aveni. A.F., ed. (1982). *Archaeoastronomy in the New World: American Primitive Astronomy*. Cambridge University Press. ISBN 0-521-24731-4.
- Aveni. A.F. (1986). "Archaeoastronomy: Past, Present and Future". *Sky and Telescope*. **72**: 456–60. Bibcode: 1986S&T....72..456A[174].
- Aveni, A.F. (1989a). *World Archaeoastronomy*. Cambridge University Press. ISBN 0-521-34180-9.
- Aveni, A.F. (1989b). *Empires of Time*. Basic Books. ISBN 0-465-01950-1.
- Aveni, A.F. (1997). *Stairways to the Stars: Skywatching in Three Great Ancient Cultures*. John Wiley and Sons. ISBN 0-471-32976-2.
- Aveni. A.F. (2003). "Archaeoastronomy in the Ancient Americas"[175] (PDF). *Journal of Archaeological Research*. **11** (2): 149–91. doi: 10.1023/A:1022971730558[176].
- Aveni, A.F. (2006). "Evidence and Intentionality: On Evidence in Archaeoastronomy". In Todd W. Bostwick; Bryan Bates. *Viewing the Sky Through Past and Present Cultures: Selected Papers from the Oxford VII International Conference on Archaeoastronomy*. Pueblo Grande Museum Anthropological Papers. **15**. City of Phoenix Parks and Recreation Department. pp. 57–70. ISBN 1-882572-38-6.
- Bahn, P. (1995). *Archaeology: A Very Short introduction*. Oxford University Press. ISBN 0-19-285379-1.
- Bauer, B. & Dearborn, D. (1995). *Astronomy and empire in the ancient Andes: the cultural origins of Inca sky watching*. University of Texas. ISBN 0-292-70837-8.
- Belmonte, J. A. (2001). "On the Orientation of Old Kingdom Egyptian Pyramids". *Archaeoastronomy: Supplement to the Journal for the History of Astronomy*. **32** (26): S1–S20. Bibcode: 2001JHAS...32....1B[177].
- Blomberg, P. (2003). "The early Hellenic Sky Map reconstructed from Archaeoastronomical and Textual Studies". In Amanda-Alice Maravelia. *Ad Astra per Aspera et per Ludum: European Archaeoastronomy and the Orientation of Munuments in the Mediterranean Basin: Papers from Session I.13, held at the European Association of Archaeologists Eighth Annual Meeting in Thessaloniki 2002*. BAR International Series 1154. Archaeopress. pp. 71–76. ISBN 1-84171-524-7.
- Bostwick, T.W. (2006). "Archaeoastronomy at the Gates of Orthodoxy: Introduction to the Oxford VII Conference on Archaeoastronomy Papers". In Todd W. Bostwick; Bryan Bates. *Viewing the Sky Through Past*

and Present Cultures: Selected Papers from the Oxford VII International Conference on Archaeoastronomy. Pueblo Grande Museum Anthropological Papers. **15**. City of Phoenix Parks and Recreation Department. pp. 1–10. ISBN 1-882572-38-6.

- Brandt, J.C. & Williamson, R.A. (1979). "The 1054 Supernova and American Rock Art". *Archaeoastronomy: Supplement to the Journal for the History of Astronomy*. **1** (10): S1–S38. Bibcode: 1979JHAS...10....1B[178].
- Broda, J. (2000). "Mesoamerican Archaeoastronomy and the Ritual Calendar". In Helaine Selin. *Astronomy Across Cultures*. Kluwer, Dordrect. pp. 225–267. ISBN 0-7923-6363-9.
- Broda, J. (2006). "Zenith Observations and the Conceptualization of Geographical Latitude in Ancient Mesoamerica: A Historical Interdisciplinary Approach". In Todd W. Bostwick; Bryan Bates. *Viewing the Sky Through Past and Present Cultures; Selected Papers from the Oxford VII International Conference on Archaeoastronomy*. Pueblo Grande Museum Anthropological Papers. **15**. City of Phoenix Parks and Recreation Department. pp. 183–212. ISBN 1-882572-38-6.
- Cairns, H.C. (2005). "Discoveries in Aboriginal Sky Mapping (Australia)". In John W. Fountain; Rolf M. Sinclair. *Current Studies in Archaeoastronomy: Conversations Across Time and Space*. Durham, North Carolina: Carolina Academic Press. pp. 523–538. ISBN 0-89089-771-9.
- Carlson, J. (Fall 1999). "Editorial: A Professor of Our Own"[179]. *Archaeoastronomy & Ethnoastronomy News* (Autumn Equinox). **33**. Retrieved 2008-03-22.
- Chamberlain, V.D. & Young, M.J. (2005). "Introduction". In Von Del Chamberlain; John Carlson & M. Jane Young. *Songs from the Sky: Indigenous Astronomical and Cosmological Traditions of the World*. Ocarina Books. pp. xi–xiv. ISBN 0-9540867-2-4.
- Chiu, B.C. & Morrison, P. (1980). "Astronomical Origin of the Offset Street Grid at Teotihuacan". *Archaeoastronomy: Supplement to the Journal for the History of Astronomy*. **11** (18): S55–S64. Bibcode: 1980JHAS...11...55C[180].
- Dearborn, D.S.P.; Seddon, M.F. & Bauer, B.S. (1998). "The Sanctuary of Titicaca: Where the Sun Returns to Earth". *Latin American Antiquity*. **9** (3): 240–258. JSTOR 971730[181]. doi: 10.2307/971730[182].
- Eogan, G. (1991). "Prehistoric and Early Historic Cultural Change at Brugh na Bóinne". *Proceedings of the Royal Irish Academy*. **91C**: 105–132.
- Fairall, A. (1999). "Precession and the layout of the Ancient Egyptian pyramids"[183]. *Astronomy & Geophysics*. The Royal Astronomical Society. **40** (4). Archived[184] from the original on 2008-02-28. Retrieved

2008-03-22.

- Fell, B. (1983). "Christian Messages in Old Irish Script Deciphered from Rock Carvings in W. VA."[185]. *Wonderful West Virginia* (47): 12–19. Archived[186] from the original on 9 May 2008. Retrieved 2008-04-27.
- Fisher, V.B. (2006). "Ignoring Archaeoastronomy: A Dying Tradition in American Archaeology". In Todd W. Bostwick; Bryan Bates. *Viewing the Sky Through Past and Present Cultures; Selected Papers from the Oxford VII International Conference on Archaeoastronomy*. Pueblo Grande Museum Anthropological Papers. **15**. City of Phoenix Parks and Recreation Department. pp. 1–10. ISBN 1-882572-38-6.
- Fountain, J. (2005). "A Database of Rock Art Solar Markers". In John W. Fountain; Rolf M. Sinclair. *Current Studies in Archaeoastronomy: Conversations Across Time and Space*. Durham, North Carolina: Carolina Academic Press. ISBN 0-89089-771-9.
- Freeth, T; Bitsakis, Y; Moussas, X; Seiradakis, JH; Tselikas, A; Mangou, H; Zafeiropoulou, M; Hadland, R; et al. (30 November 2006). "Decoding the ancient Greek astronomical calculator known as the Antikythera Mechanism". *Nature*. **444** (7119): 587–591. Bibcode: 2006Natur.444..587F[187]. PMID 17136087[188]. doi: 10.1038/nature05357[189].
- Gallagher, I.J. (1983). "Light Dawns on West Virginia History"[190]. *Wonderful West Virginia* (47): 7–11. Archived[191] from the original on 11 May 2008. Retrieved 2008-04-27.
- Gingerich, O. (24 March 2000). "Stone and star gazing"[192]. *Times Higher Education Supplement*: 24. Retrieved 2008-03-22.
- Gummerman, G.J. & Warburton, M (2005). "The Universe in Cultural Context: An Essay". In John W. Fountain & Rolf M. Sinclair. *Current Studies in Archaeoastronomy: Conversations Across Time and Space*. Durham, North Carolina: Carolina Academic Press. ISBN 0-89089-771-9.
- Hamacher, D.W. (2012). *On the Astronomical Knowledge and Traditions of Aboriginal Australians*. PhD Thesis: Macquarie University, Sydney, Australia.
- Hancock, G (1996). *Fingerprints of the Gods*. New York: Three Rivers Press. ISBN 0-517-88729-0.
- Hannah, R (1994). "The Constellations on Achilles' Shield (Iliad 18. 485–489)"[193]. *Electronic Antiquity*. **II.4**. Archived[194] from the original on 21 April 2008. Retrieved 2008-03-22.
- Hawkins, G (1976). *Stonehenge Decoded*. Fontana. ISBN 0-00-632315-4.
- Heggie, D.C. (1982). *Archaeoastronomy in the Old World*. CUP. ISBN 0-521-24734-9.

- Hicks, R. (Fall 1993). "Beyond Alignments"[195]. *Archaeoastronomy & Ethnoastronomy News* (September Equinox). **9**. Retrieved 2008-03-22.
- Hoskin, M (1996). "Book Reviews: Proceedings of Oxford 3". *Archaeoastronomy: Supplement to the Journal for the History of Astronomy*. **21** (27): S85–S87. Bibcode: 1996JHAS...27...85H[196].
- Hoskin, M (1999). *The Cambridge Concise History of Astronomy*. CUP. ISBN 0-521-57600-8.
- Hoskin, M. (2001). *Tombs, Temples, and Their Orientations: A New Perspective on Mediterranean Prehistory*. Ocarina Books. ISBN 0-9540867-1-6.
- Hugh-Jones, Stephen (1982). "The Pleiades and Scorpius in Barasana Cosmology". In Anthony F. Aveni; Gary Urton. *Ethnoastronomy and Archaeoastronomy in the American Tropics*. Annals of the New York Academy of Sciences. **385**. New York, New York: New York Academy of Sciences. pp. 183–201. Bibcode: 1982NYASA.385..183H[197]. ISBN 0-89766-160-5. doi: 10.1111/j.1749-6632.1982.tb34265.x[198].
- Isager, S. & Skydsgaard, J.E. (1992). *Ancient Greek Agriculture*. Routledge. ISBN 0-415-00164-1.
- Iwaniszewski, S. (Winter 1995). "Alignments and Orientations Again"[199]. *Archaeoastronomy & Ethnoastronomy News* (December Solstice). **18**. Retrieved 2008-03-22.
- Iwaniszewski, S. (2001). "Time and space in social systems - further issues for theoretical archaeoastronomy". In Clive Ruggles; Frank Prendergast; Tom Ray. *Astronomy, Cosmology and Landscape: Proceedings of the SEAC 98 Meeting, Dublin, Ireland*. Ocarina Books. pp. 1–7. ISBN 0-9540867-0-8.
- Iwaniszewski, S. (2003). "The erratic ways of studying astronomy in culture". In Mary Blomberg; Peter E. Blomberg; Göran Henriksson. *Calendars, Symbols and Orientations: Legacies of Astronomy in Culture (Proceedings of the 9th annual meeting of the European Society from Astronomy in Culture (SEAC), Stockholm, 27–30 August 2001*. Uppsala. pp. 7–10. ISBN 91-506-1674-9.
- Johnson, W. (1912). *Byways of British Archaeology*. Cambridge University Press.
- Kelley, D.H. & Milone, E.F. (2005). *Exploring Ancient Skies: An Encyclopedic Survey of Archaeoastronomy*. Springer-Verlag. ISBN 0-387-95310-8.
- Kintigh, K. (Fall 1992). "I wasn't going to say anything, but since you asked: Archaeoastronomy and Archaeology"[200]. *Archaeoastronomy & Ethnoastronomy News* (September Equinox). **5**. Retrieved 2008-03-22.
- Krupp, E.C. (1979). *In Search of Ancient Astronomies*. Chatto and Windus. ISBN 0-7011-2314-1.

- Krupp, E.C. (1988). "Light in the Temples". In C.L.N. Ruggles. *Records in Stone: Papers in Memory of Alexander Thom*. CUP. pp. 473–499. ISBN 0-521-33381-4.
- Krupp, E.C. (1997a). *Skywatchers, Shamans and Kings*. John Wiley and Sons. ISBN 0-471-32975-4.
- Krupp, E.C. (February 1997). "Rambling Through the Skies: Pyramid Marketing SchemesPyramid Marketing Schemes"[201]. *Sky and Telescope*. **94** (2): 64.
- Krupp, E. C.; Billo, Evelyn; Mark, Robert (2010). "Star Trek: Recovery and Review of the First Alleged Supernova Rock Art". *Archaeoastronomy: The Journal of Astronomy in Culture*. University of Texas Press. **23**: 35–43.
- Lesser, W.H. (1983). "Cult Archaeology Strikes Again: A Case for Pre-Columbian Irishmen in the Mountain State?"[202]. *West Virginia Archeologist*. **35**: 48–52.
- McCluskey, S.C. (2000). "The Inconstant Moon: Lunar Astronomies in Different Cultures". *Archaeoastronomy: The Journal of Astronomy in Culture*. **15**: 14–31. Bibcode: 2000Arch...15...14M[203].
- McCluskey, S.C. (2001). "Etnoscienza dei Pueblo". In Sandro Petruccioli. *Storia della Scienza, vol. 2, Cina, India, Americhe, Sec. 3, "Le Civilta Precolombiane"*. Rome: Istituto della Enciclopedia Italiana. pp. 1002–1009.
- McCluskey, S.C. (2004). "The Study of Astronomies in Cultures as Reflected in Dissertations and Theses". *Archaeoastronomy: The Journal of Astronomy in Culture*. **16**: 20–25.
- McCluskey, S.C. (2005). "Different Astronomies, Different Cultures and the Question of Cultural Relativism". In John W. Fountain; Rolf M. Sinclair. *Current Studies in Archaeoastronomy: Conversations Across Time and Space*. Carolina Academic Press. pp. 69–79. ISBN 0-89089-771-9.
- MacKie, E (1977). *Science and Society in Prehistoric Britain*. Paul Elek. ISBN 0-236-40041-X.
- MacKie, E (1997). "Maeshowe and the winter solstice: ceremonial aspects of the Orkney Grooved Ware culture". *Antiquity*. **71** (272): 338–359. doi: 10.1017/S0003598X00084969[204].
- MacKie, E (2006). "New Evidence for a Professional Priesthood in the European Early Bronze Age". In Todd W. Bostwick; Bryan Bates. *Viewing the Sky Through Past and Present Cultures: Selected Papers from the Oxford VII International Conference on Archaeoastronomy*. Pueblo Grande Museum Anthropological Papers. **15**. City of Phoenix Parks and Recreation Department. pp. 343–362. ISBN 1-882572-38-6.
- MacKie, E. W. (2009). "The Prehistoric Solar Calendar: An Out of Fashion Idea Revisited with New Evidence". *Time and Mind*. **2** (1): 9–46.

doi: 10.2752/175169709x374263[205].

- Magli, G. (2008). "On the Orientation of Roman Towns in Italy"[206]. *Oxford Journal of Archaeology*. **27** (1): 63–71. doi: 10.1111/j.1468-0092.2007.00296.x[207].
- Magli, G. (2013). *Architecture, Astronomy and Sacred Landscape in Ancient Egypt*. CUP.
- Magli, G. (2015). *Archaeoastronomy. Introduction to the science of stars and stones*. Springer, NY.
- Meller, H. (January 2004). "Star search". *National Geographic*: 76–8.
- Michell, J. (2001). *A Little History of Astro-Archaeology*. Thames & Hudson. ISBN 0-500-27557-2.
- Milbraith, S. (1988). "Astronomical Images and Orientations in the Architecture of Chichen Itzá". In A.F. Aveni. *New Directions in American Archaeoastronomy*. BAR International Series. **454**. BAR. pp. 54–79. ISBN 0-86054-583-0.
- Milbraith, S. (1999). *Star Gods of the Maya: Astronomy in Art, Folklore and Calendars*. University of Texas Press. ISBN 0-292-75226-1.
- Neugebauer, Otto (1980). "On the Orientation of Pyramids". *Centaurus*. **24**: 1–3. Bibcode: 1980Cent...24....1N[208]. doi: 10.1111/j.1600-0498.1980.tb00362.x[209].
- O'Kelly, M.J. (1982). *Newgrange: Archaeology, Art and Legend*. Thames and Hudson. ISBN 0-500-39015-0.
- Parker Pearson, M (2007). "The age of Stonehenge"[210]. *Antiquity*. **81** (313): 617–639. doi: 10.1017/s0003598x00095624[211].
- Pedersen, Olaf (1982). "The Present Position of Archaeo-Astronomy". In D. C. Heggie. *Archaeoastronomy in the Old World*. Cambridge: Cambridge University Press. pp. 265–274. ISBN 0-521-24734-9.
- Pingree, D. (1982). "Hellenophilia versus the History of Science". *Isis*. **83** (4): 554–563. Bibcode: 1992Isis...83..554P[212]. doi: 10.1086/356288[213].. reprinted in Michael H. Shank, ed., *The Scientific Enterprise in Antiquity and the Middle Ages* (Chicago: Univ. of Chicago Pr., 2000), pp. 30–39.
- Poss, R.L. (2005). "Interpreting Rock Art: European and Anasazi Representations of Spirituality". In John W. Fountain; Rolf M. Sinclair. *'Current Studies in Archaeoastronomy: Conversations Across Time and Space*. Durham, North Carolina: Carolina Academic Press. pp. 81–98. ISBN 0-89089-771-9.
- Preston R.A. & Preston A.L. (2005). "Consistent Forms of Solstice Sunlight Interaction with Petroglyphs throughout the Prehistoric American Southwest". In John W. Fountain & Rolf M. Sinclair. *Current Studies in Archaeoastronomy: Conversations Across Time and Space*. Durham, North Carolina: Carolina Academic Press. ISBN 0-89089-771-9.
- Pyle, R.L. (1983). "A Message from the Past"[214]. *Wonderful West*

Virginia (47): 3–6. Archived[215] from the original on 11 May 2008. Retrieved 2008-04-27.

- Robins M. & Ewing E. (1989). "The Sun is in His House: Summer Solstice at San Carlos Mesa". In Ken Hedges. *Rock Art Papers, vol. 6*. San Diego Museum Papers. **24**. San Diego Museum.
- Rufinus. "The destruction of the Serapeum"[216]. Archived[217] from the original on 6 April 2008. Retrieved 2008-03-22.
- Ruggles, C.L.N. (1993). *Archaeoastronomy in the 1990s*. Group D Publications. ISBN 1-874152-01-2.
- Ruggles, C.L.N. (1999). *Astronomy in Prehistoric Britain and Ireland*. Yale University Press. ISBN 0-300-07814-5.
- Ruggles, C.L.N. (2000). "Ancient Astronomies - Ancient Worlds". *Archaeoastronomy: Supplement to the Journal for the History of Astronomy*. **25** (31): S65–S76. Bibcode: 2000ArchS..25...65R[218].
- Ruggles, C.L.N. (2005). *Ancient Astronomy*. ABC-Clio. ISBN 1-85109-477-6.
- Ruggles, Clive L. N., ed. (2014). *The Handbook of Archaeoastronomy and Ethnoastronomy*. New York: Springer. ISBN 978-1-4614-6140-1. doi: 10.1007/978-1-4614-6141-8[219]. Three volumes; 217 articles.
- Ruggles, C.L.N. & Barclay, G. (2000). "Cosmology, calendars and society in Neolithic Orkney: a rejoinder to Euan MacKie". *Antiquity*. **74** (283): 62–74. doi: 10.1017/S0003598X00066151[220].
- Ruggles, C.L.N.; Cotte, M., eds. (2010). *Heritage Sites of Astronomy and Archaeoastronomy in the context of the UNESCO World Heritage Convention: A Thematic Study*[221]. Paris: ICOMOS / IAU. ISBN 978-2-918086-01-7.
- Ruggles, C.L.N. & Saunders, N.J. (1993). "The Study of Cultural Astronomy". In Clive L.N. Ruggles & Nicholas J. Saunders. *Astronomies and Cultures*. University Press of Colorado. pp. 1–31. ISBN 0-87081-319-6.
- Saethre, E. (2007). "Close encounters: UFO beliefs in a remote Australian Aboriginal community". *Journal of the Royal Anthropological Institute*. **13** (4): 901–915. doi: 10.1111/j.1467-9655.2007.00463.x[222].
- Salt, A. & Boutsikas, E. (2005). "Knowing when to consult the oracle at Delphi". *Antiquity*. **79** (305): 562–72. doi: 10.1017/S0003598X00114504[223].
- Schaefer, B.E. (2002). "The Great Ptolemy-Hipparchus Dispute". *Sky and Telescope*. **103** (2): 38–66. Bibcode: 2002S&T...103b..38S[224].
- Schaefer, B.E. (2006a). "Case Studies of Three of the Most Famous Claimed Archaeoastronomical Alignments in North America". In Todd W. Bostwick; Bryan Bates. *Viewing the Sky Through Past and Present Cultures: Selected Papers from the Oxford VII International Conference on Archaeoastronomy*. Pueblo Grande Museum Anthropological Pa-

pers. **15**. City of Phoenix Parks and Recreation Department. pp. 27–56. ISBN 1-882572-38-6.

- Schaefer, B.E. (2006b). "No Astronomical Alignments at the Caracol". In Todd W. Bostwick; Bryan Bates. *Viewing the Sky Through Past and Present Cultures: Selected Papers from the Oxford VII International Conference on Archaeoastronomy*. Pueblo Grande Museum Anthropological Papers. **15**. City of Phoenix Parks and Recreation Department. pp. 71–77. ISBN 1-882572-38-6.
- Schaefer, B. E.; Liller, W. (1990). "Refraction near the Horizon". *Publications of the Astronomical Society of the Pacific*. **102**: 796–805. Bibcode: 1990PASP..102..796S[225]. doi: 10.1086/132705[226].
- Schlosser, W. (2002). "Zur astronomischen Deuteung der Himmelschreibe vom Nebra". *Archäologie in Sachsen-Anhalt*. **1/02**: 21–3.
- Sinclair, R.M. (2005). "The Nature of Archaeoastronomy". In John W. Fountain; Rolf M. Sinclair. *Current Studies in Archaeoastronomy: Conversations Across Time and Space*. Carolina Academic Press. pp. 3–13. ISBN 0-89089-771-9.
- Sinclair, R.M. (2006). "The Nature of Archaeoastronomy"". In Todd W. Bostwick; Bryan Bates. *Viewing the Sky Through Past and Present Cultures; Selected Papers from the Oxford VII International Conference on Archaeoastronomy*. Pueblo Grande Museum Anthropological Papers. **15**. City of Phoenix Parks and Recreation Department. pp. 13–26. ISBN 1-882572-38-6.
- Sofaer, A., ed. (2008). *Chaco Astronomy: An Ancient American Cosmology*. Santa Fe, New Mexico: Ocean Tree Books. ISBN 978-0-943734-46-0.
- Steel, D. (June 1999). "Stonehenge and the Terror in the Sky"[227]. *British Archaeology*. **45**. Retrieved 2008-09-06.
- Spence, K (16 November 2000). "Ancient Egyptian Chronoology and the astronomical orientation of the pyramids". *Nature*. **408** (6810): 320–324. Bibcode: 2000Natur.408..320S[228]. PMID 11099032[229]. doi: 10.1038/35042510[230].
- Šprajc, I. (2015). Governor's Palace at Uxmal. In: *Handbook of Archaeoastronomy and Ethnoastronomy*, ed. by Clive L. N. Ruggles, New York: Springer, pp. 773–781
- Šprajc, I., and P. F. Sánchez Nava (2013). Astronomía en la arquitectura de Chichén Itzá: una reevaluación[231]. *Estudios de Cultura Maya* XLI: 31–60.
- Thom, A. (1967). *Megalithic Sites in Britain*. Oxford: Clarendon Press. ISBN 0-19-813148-8.
- Thom. A.S. (1988). "A personal note about my late father, Alexander Thom". In C.L.N. Ruggles. *Records in Stone: Papers in Memory of*

Alexander Thom. CUP. pp. 3–13. ISBN 0-521-33381-4.

- Trotter, A.P. (1927). "Stonehenge as an Astronomical Instrument". *Antiquity*. **1** (1): 42–53. doi: 10.1017/S0003598X00000053[232].
- Turton, D. & Ruggles, C.L.N. (1978). "Agreeing to Disagree: The Measurement of Duration in a Southwestern Ethiopian Community". *Current Anthropology*. **19.3** (3): 585–600. doi: 10.1086/202140[233].
- Urton, G. (1981). *At the crossroads of the earth and the sky: an Andean cosmology*. University of Texas. ISBN 0-292-70349-X.
- van Driel-Murray, C. (2002). "Regarding the Stars". In M Carruthers; C. van Driel-Murray; A. Gardner; J. Lucas; et al. *TRAC 2001: Proceedings of the Eleventh Annual Theoretical Roman Archaeology Conference Glasgow 2001*. Oxbow Books. pp. 96–103. ISBN 1-84217-075-9.
- Wise, R.B. (2003). "The Observations of the 2002 Winter Solstice at Luther Elkins Petroglyph (46 Wm 3)"[234]. Council for West Virginia Archaeology. Retrieved 2008-03-22.
- Williamson, Ray A. (1987). "Light and Shadow, Ritual, and Astronomy in Anasazi Structures". In John B. Carlson; W. James Judge. *Astronomy and Ceremony in the Prehistoric Southwest*. Papers of the Maxwell Museum of Anthropology. **2**. Albuquerque, NM. pp. 71–88. ISBN 0-912535-03-2.
- Witzel, M. (May 2001). "Autochthonous Aryans? The Evidence from Old Indian and Iranian Text"[235]. *Electronic Journal of Vedic Studies*. **7** (3): §28–30. Archived from the original[236] on 2008-03-28.
- Xu, Z.; Pankenier, D.W. & Jiang, Y. (2000). *East Asian Archaeoastronomy: Historical Records of Astronomical Observations of China, Japan and Korea*. Amsterdam: Gordon & Breach Science Publ. ISBN 90-5699-302-X.
- Young, M.J. (2005). "Ethnoastronomy and the Problem of Interpretation: A Zuni Example". In Von Del Chamberlain; John Carlson; M. Jane Young. *Songs from the Sky: Indigenous and Cosmological Traditions of the World*. Ocarina Books. ISBN 0-9540867-2-4.
- Zeilik, M. (1985). "The Ethnoastronomy of the Historic Pueblos, I: Calendrical Sun Watching". *Archaeoastronomy: Supplement to the Journal for the History of Astronomy*. **8** (16): S1–S24. Bibcode: 1985JHAS...16....1Z[237].
- Zeilik, M. (1986). "The Ethnoastronomy of the Historic Pueblos, II: Moon Watching". *Archaeoastronomy: Supplement to the Journal for the History of Astronomy*. **10** (17): S1–S22. Bibcode: 1986JHAS...17....1Z[238].

Further reading

- Ruggles, Clive L. N., ed. (2014). *The Handbook of Archaeoastronomy and Ethnoastronomy*. New York: Springer. ISBN 978-1-4614-6140-1. doi: 10.1007/978-1-4614-6141-8[219]. Three volumes; 217 articles.
- Magli, G. (2015). *Archaeoastronomy. Introduction to the science of stars and stones*. Springer, NY.

External links

Look up ***archaeoastronomy***in Wiktionary, the free dictionary.

- Astronomy before History, by Clive Ruggles and Michael Hoskins[239], a chapter from *The Cambridge Concise History of Astronomy,* Michael Hoskin ed., 1999
- Clive Ruggles's webpage:[240] images, bibliography, software, and synopsis of his course at the University of Leicester
- Home page of the free Archaeoastronomy course online at the Politecnico of Milan[241] Giulio Magli's MOOC -Archaeoastronomy. The science of stars and stones.
- Traditions of the Sun[242] — NASA and others exploring the world's ancient observatories.
- Ancient Observatories: Timeless Knowledge[243] NASA Poster on ancient (and modern) observatories.

Societies

- ISAAC[166], The International Society for Archaeoastronomy and Astronomy in Culture.
- SEAC[167] La Société Européenne pour l'Astronomie dans la Culture. Site in English.
- SIAC[244] La Sociedad Interamericana de Astronomía en la Cultura.
- Society for the History of Astronomy[245]

Journals

- Archaeoastronomy and Ethnoastronomy News[246]
- Archaeoastronomy: Supplement to the Journal for the History of Astronomy[247]
- Archaeoastronomy: The Journal of Astronomy in Culture[248]
- Culture and Cosmos[249]
- Journal for the History of Astronomy[250]

- Archaeoastronomy and Ancient Technologies (open access)[251]

<indicator name=”good-star”> ⊕ </indicator>

Ancient Civilizations

Babylonian astronomy

According to Asger Aaboe, the origins of Western astronomy can be found in Mesopotamia, and all Western efforts in the exact sciences are descendants in direct line from the work of the late Babylonian astronomers.[252] Modern knowledge of Sumerian astronomy is indirect, via the earliest Babylonian star catalogues dating from about 1200 BC. The fact that many star names appear in Sumerian suggests a continuity reaching into the Early Bronze Age.

The history of astronomy in Mesopotamia, and the world, begins with the Sumerians who developed the earliest writing system—known as cuneiform—around 3500–3200 BC. The Sumerians developed a form of astronomy that had an important influence on the sophisticated astronomy of the Babylonians. Astrolatry, which gave planetary gods an important role in Mesopotamian mythology and religion, began with the Sumerians. They also used a sexagesimal (base 60) place-value number system, which simplified the task of recording very great and very small numbers. The modern practice of dividing a circle into 360 degrees, of 60 minutes each hour, began with the Sumerians.

During the 8th and 7th centuries BC, Babylonian astronomers developed a new empirical approach to astronomy. They began studying philosophy dealing with the ideal nature of the universe and began employing an internal logic within their predictive planetary systems. This was an important contribution to astronomy and the philosophy of science, and some scholars have thus referred to this new approach as the first scientific revolution.[253] This new approach to astronomy was adopted and further developed in Greek and Hellenistic astronomy. Classical Greek and Latin sources frequently use the term Chaldeans for the astronomers of Mesopotamia, who were, in reality, priest-scribes specializing in astrology and other forms of divination.

Figure 21: *Babylonian tablet recording Halley's comet in 164 BC.*

Only fragments of Babylonian astronomy have survived, consisting largely of contemporary clay tablets with ephemerides and procedure texts, hence current knowledge of Babylonian planetary theory is in a fragmentary state. Nevertheless, the surviving fragments show that, according to the historian A. Aaboe, Babylonian astronomy was "the first and highly successful attempt at giving a refined mathematical description of astronomical phenomena" and that "all subsequent varieties of scientific astronomy, in the Hellenistic world, in India, in Islam, and in the West—if not indeed all subsequent endeavour in the exact sciences—depend upon Babylonian astronomy in decisive and fundamental ways."

Old Babylonian astronomy

Old Babylonian astronomy refers to the astronomy that was practiced during and after the First Babylonian Dynasty (ca. 1830 BC) and before the Neo-Babylonian Empire (ca. 626 BC).

The Babylonians were the first to recognize that astronomical phenomena are periodic and apply mathematics to their predictions. Tablets dating back to the Old Babylonian period document the application of mathematics to the variation in the length of daylight over a solar year. Centuries of Babylonian observations of celestial phenomena are recorded in the series of cuneiform tablets known as the *Enûma Anu Enlil*—the oldest significant astronomical text that we possess is Tablet 63 of the *Enûma Anu Enlil*, the Venus tablet of Ammisaduqa, which lists the first and last visible risings of Venus over a period of about 21 years. It is the earliest evidence that planetary phenomena were recognized as periodic.

The MUL.APIN contains catalogues of stars and constellations as well as schemes for predicting heliacal risings and settings of the planets, and lengths of daylight as measured by a water clock, gnomon, shadows, and intercalations. The Babylonian GU text arranges stars in 'strings' that lie along declination circles and thus measure right-ascensions or time intervals, and also employs the stars of the zenith, which are also separated by given right-ascensional differences. There are dozens of cuneiform Mesopotamian texts with real observations of eclipses, mainly from Babylonia.

Planetary theory

The Babylonians were the first civilization known to possess a functional theory of the planets. The oldest surviving planetary astronomical text is the Babylonian Venus tablet of Ammisaduqa, a 7th-century BC copy of a list of observations of the motions of the planet Venus that probably dates as early as the second millennium BC. The Babylonian astrologers also laid the foundations of what would eventually become Western astrology. The *Enuma anu enlil*, written during the Neo-Assyrian period in the 7th century BC, comprises a list of omens and their relationships with various celestial phenomena including the motions of the planets.

Cosmology

In contrast to the world view presented in Mesopotamian and Assyro-Babylonian literature, particularly in Mesopotamian and Babylonian mythology, very little is known about the cosmology and world view of the ancient Babylonian astrologers and astronomers. This is largely due to the current fragmentary state of Babylonian planetary theory, and also due to Babylonian astronomy being independent from cosmology at the time. Nevertheless, traces of cosmology can be found in Babylonian literature and mythology.

In Babylonian cosmology, the Earth and the heavens were depicted as a "spatial whole, even one of round shape" with references to "the circumference of heaven and earth" and "the totality of heaven and earth". Their worldview was not exactly geocentric either. The idea of geocentrism, where the center of the Earth is the exact center of the universe, did not yet exist in Babylonian cosmology, but was established later by the Greek philosopher Aristotle's *On the Heavens*. In contrast, Babylonian cosmology suggested that the cosmos revolved around circularly with the heavens and the earth being equal and joined as a whole. The Babylonians and their predecessors, the Sumerians, also believed in a plurality of heavens and earths. This idea dates back to Sumerian incantations of the 2nd millennium BC, which refers to there being seven heavens and seven earths, linked possibly chronologically to the creation by seven generations of gods.

Neo-Babylonian astronomy

Neo-Babylonian astronomy refers to the astronomy developed by Chaldean astronomers during the Neo-Babylonian, Achaemenid, Seleucid, and Parthian periods of Mesopotamian history. A significant increase in the quality and frequency of Babylonian observations appeared during the reign of Nabonassar (747–734 BC). The systematic records of ominous phenomena in Babylonian astronomical diaries that began at this time allowed for the discovery of a repeating 18-year Saros cycle of lunar eclipses, for example. The Greco-Egyptian astronomer Ptolemy later used Nabonassar's reign to fix the beginning of an era, since he felt that the earliest usable observations began at this time.

The last stages in the development of Babylonian astronomy took place during the time of the Seleucid Empire (323–60 BC). In the 3rd century BC, astronomers began to use "goal-year texts" to predict the motions of the planets. These texts compiled records of past observations to find repeating occurrences of ominous phenomena for each planet. About the same time, or shortly afterwards, astronomers created mathematical models that allowed them to predict these phenomena directly, without consulting past records.

Arithmetical and geometrical methods

Though there is a lack of surviving material on Babylonian planetary theory, it appears most of the Chaldean astronomers were concerned mainly with ephemerides and not with theory. It had been thought that most of the predictive Babylonian planetary models that have survived were usually strictly empirical and arithmetical, and usually did not involve geometry, cosmology, or speculative philosophy like that of the later Hellenistic models, though the Babylonian astronomers were concerned with the philosophy dealing with the ideal nature of the early universe. Babylonian procedure texts describe, and ephemerides employ, arithmetical procedures to compute the time and place of significant astronomical events. More recent analysis of previously unpublished cuneiform tablets in the British Museum, dated between 350 and 50 BC, demonstrates that Babylonian astronomers sometimes used geometrical methods, prefiguring the methods of the Oxford Calculators, to describe the motion of Jupiter over time in an abstract mathematical space.

In contrast to Greek astronomy which was dependent upon cosmology, Babylonian astronomy was independent from cosmology. Whereas Greek astronomers expressed "prejudice in favor of circles or spheres rotating with uniform motion", such a preference did not exist for Babylonian astronomers, for whom uniform circular motion was never a requirement for planetary orbits. There is no evidence that the celestial bodies moved in uniform circular motion, or along celestial spheres, in Babylonian astronomy.

Contributions made by the Chaldean astronomers during this period include the discovery of eclipse cycles and saros cycles, and many accurate astronomical observations. For example, they observed that the Sun's motion along the ecliptic was not uniform, though they were unaware of why this was; it is today known that this is due to the Earth moving in an elliptic orbit around the Sun, with the Earth moving swifter when it is nearer to the Sun at perihelion and moving slower when it is farther away at aphelion.

Chaldean astronomers known to have followed this model include Naburimannu (fl. 6th–3rd century BC), Kidinnu (d. 330 BC), Berossus (3rd century BCE), and Sudines (fl. 240 BCE). They are known to have had a significant influence on the Greek astronomer Hipparchus and the Egyptian astronomer Ptolemy, as well as other Hellenistic astronomers.

Heliocentric astronomy

The only surviving planetary model from among the Chaldean astronomers is that of Seleucus of Seleucia (b. 190 BC), who supported Aristarchus of Samos' heliocentric model.[254] Seleucus is known from the writings of Plutarch, Aetius, Strabo, and Muhammad ibn Zakariya al-Razi. Strabo lists Seleucus as one of the four most influential Chaldean/Babylonian astronomers, alongside Kidenas (Kidinnu), Naburianos (Naburimannu), and Sudines. Their works were originally written in the Akkadian language and later translated into Greek. Seleucus, however, was unique among them in that he was the only one known to have supported the heliocentric theory of planetary motion proposed by Aristarchus,[255,256,257] where the Earth rotated around its own axis which in turn revolved around the Sun. According to Plutarch, Seleucus even proved the heliocentric system through reasoning, though it is not known what arguments he used.

According to Lucio Russo, his arguments were probably related to the phenomenon of tides.[258] Seleucus correctly theorized that tides were caused by the Moon, although he believed that the interaction was mediated by the Earth's atmosphere. He noted that the tides varied in time and strength in different parts of the world. According to Strabo (1.1.9), Seleucus was the first to state that the tides are due to the attraction of the Moon, and that the height of the tides depends on the Moon's position relative to the Sun.

According to Bartel Leendert van der Waerden, Seleucus may have proved the heliocentric theory by determining the constants of a geometric model for the heliocentric theory and by developing methods to compute planetary positions using this model. He may have used trigonometric methods that were available in his time, as he was a contemporary of Hipparchus.

None of his original writings or Greek translations have survived, though a fragment of his work has survived only in Arabic translation, which was later referred to by the Persian philosopher Muhammad ibn Zakariya al-Razi (865-925).

Babylonian influence on Hellenistic astronomy

Many of the works of ancient Greek and Hellenistic writers (including mathematicians, astronomers, and geographers) have been preserved up to the present time, or some aspects of their work and thought are still known through later references. However, achievements in these fields by earlier ancient Near Eastern civilizations, notably those in Babylonia, were forgotten for a long time. Since the discovery of key archaeological sites in the 19th century, many cuneiform writings on clay tablets have been found, some of them related to astronomy. Most known astronomical tablets have been described by Abraham Sachs and later published by Otto Neugebauer in the *Astronomical Cuneiform Texts* (*ACT*).

Since the rediscovery of the Babylonian civilization, it has become apparent that Hellenistic astronomy was strongly influenced by the Chaldeans. The best documented borrowings are those of Hipparchus (2nd century BCE) and Claudius Ptolemy (2nd century CE).

Early influence

Many scholars agree that the Metonic cycle is likely to have been learned by the Greeks from Babylonian scribes. Meton of Athens, a Greek astronomer of the 5th century BCE, developed a lunisolar calendar based on the fact that 19 solar years is about equal to 235 lunar months, a period relation already known to the Babylonians.

In the 4th century, Eudoxus of Cnidus wrote a book on the fixed stars. His descriptions of many constellations, especially the twelve signs of the zodiac, are suspiciously similar to Babylonian originals. The following century Aristarchus of Samos used an eclipse cycle of Babylonian origin called the

Saros cycle to determine the year length. However, all these examples of early influence must be inferred and the chain of transmission is not known.

Influence on Hipparchus and Ptolemy

In 1900, Franz Xaver Kugler demonstrated that Ptolemy had stated in his *Almagest* IV.2 that Hipparchus improved the values for the Moon's periods known to him from "even more ancient astronomers" by comparing eclipse observations made earlier by "the Chaldeans", and by himself. However Kugler found that the periods that Ptolemy attributes to Hipparchus had already been used in Babylonian ephemerides, specifically the collection of texts nowadays called "**System B**" (sometimes attributed to Kidinnu). Apparently Hipparchus only confirmed the validity of the periods he learned from the Chaldeans by his newer observations. Later Greek knowledge of this specific Babylonian theory is confirmed by 2nd-century papyrus, which contains 32 lines of a single column of calculations for the Moon using this same "System B", but written in Greek on papyrus rather than in cuneiform on clay tablets.[259]

It is clear that Hipparchus (and Ptolemy after him) had an essentially complete list of eclipse observations covering many centuries. Most likely these had been compiled from the "diary" tablets: these are clay tablets recording all relevant observations that the Chaldeans routinely made. Preserved examples date from 652 BC to AD 130, but probably the records went back as far as the reign of the Babylonian king Nabonassar: Ptolemy starts his chronology with the first day in the Egyptian calendar of the first year of Nabonassar; i.e., 26 February 747 BC.

This raw material by itself must have been tough to use, and no doubt the Chaldeans themselves compiled extracts of e.g., all observed eclipses (some tablets with a list of all eclipses in a period of time covering a saros have been found). This allowed them to recognise periodic recurrences of events. Among others they used in System B (cf. *Almagest* IV.2):

- 223 (synodic) months = 239 returns in anomaly (anomalistic month) = 242 returns in latitude (draconic month). This is now known as the saros period which is very useful for predicting eclipses.
- 251 (synodic) months = 269 returns in anomaly
- 5458 (synodic) months = 5923 returns in latitude
- 1 synodic month = 29;31:50:08:20 days (sexagesimal; 29.53059413... days in decimals = 29 days 12 hours 44 min 3⅓ s)

The Babylonians expressed all periods in synodic months, probably because they used a lunisolar calendar. Various relations with yearly phenomena led to different values for the length of the year.

Similarly various relations between the periods of the planets were known. The relations that Ptolemy attributes to Hipparchus in *Almagest* IX.3 had all already been used in predictions found on Babylonian clay tablets.

Other traces of Babylonian practice in Hipparchus' work are

- first Greek known to divide the circle in 360 degrees of 60 arc minutes.
- first consistent use of the sexagesimal number system.
- the use of the unit *pechus* ("cubit") of about 2° or 2½°.
- use of a short period of 248 days = 9 anomalistic months.

Means of transmission

All this knowledge was transferred to the Greeks probably shortly after the conquest by Alexander the Great (331 BC). According to the late classical philosopher Simplicius (early 6th century), Alexander ordered the translation of the historical astronomical records under supervision of his chronicler Callisthenes of Olynthus, who sent it to his uncle Aristotle. It is worth mentioning here that although Simplicius is a very late source, his account may be reliable. He spent some time in exile at the Sassanid (Persian) court, and may have accessed sources otherwise lost in the West. It is striking that he mentions the title *tèresis* (Greek: guard) which is an odd name for a historical work, but is in fact an adequate translation of the Babylonian title *massartu* meaning "guarding" but also "observing". Anyway, Aristotle's pupil Callippus of Cyzicus introduced his 76-year cycle, which improved upon the 19-year Metonic cycle, about that time. He had the first year of his first cycle start at the summer solstice of 28 June 330 BC (Julian proleptic date), but later he seems to have counted lunar months from the first month after Alexander's decisive battle at Gaugamela in fall 331 BC. So Callippus may have obtained his data from Babylonian sources and his calendar may have been anticipated by Kidinnu. Also it is known that the Babylonian priest known as Berossus wrote around 281 BC a book in Greek on the (rather mythological) history of Babylonia, the *Babyloniaca*, for the new ruler Antiochus I; it is said that later he founded a school of astrology on the Greek island of Kos. Another candidate for teaching the Greeks about Babylonian astronomy/astrology was Sudines who was at the court of Attalus I Soter late in the 3rd century BC.

In any case, the translation of the astronomical records required profound knowledge of the cuneiform script, the language, and the procedures, so it seems likely that it was done by some unidentified Chaldeans. Now, the Babylonians dated their observations in their lunisolar calendar, in which months and years have varying lengths (29 or 30 days; 12 or 13 months respectively). At the time they did not use a regular calendar (such as based on the Metonic cycle like they did later), but started a new month based on observations of the

New Moon. This made it very tedious to compute the time interval between events.

What Hipparchus may have done is transform these records to the Egyptian calendar, which uses a fixed year of always 365 days (consisting of 12 months of 30 days and 5 extra days): this makes computing time intervals much easier. Ptolemy dated all observations in this calendar. He also writes that "All that he (=Hipparchus) did was to make a compilation of the planetary observations arranged in a more useful way" (*Almagest* IX.2). Pliny states (*Naturalis Historia* II.IX(53)) on eclipse predictions: "After their time (=Thales) the courses of both stars (=Sun and Moon) for 600 years were prophesied by Hipparchus, ...". This seems to imply that Hipparchus predicted eclipses for a period of 600 years, but considering the enormous amount of computation required, this is very unlikely. Rather, Hipparchus would have made a list of all eclipses from Nabonasser's time to his own.

References

Wikimedia Commons has media related to ***Babylonian astronomy***.

- Aaboe, Asger. *Episodes from the Early History of Astronomy*. New York: Springer, 2001. ISBN 0-387-95136-9
- Jones, Alexander. "The Adaptation of Babylonian Methods in Greek Numerical Astronomy." *Isis*, 82(1991): 441-453; reprinted in Michael Shank, ed. *The Scientific Enterprise in Antiquity and the Middle Ages*. Chicago: Univ. of Chicago Pr., 2000. ISBN 0-226-74951-7
- Kugler, F. X. *Die Babylonische Mondrechnung* ("The Babylonian lunar computation.") Freiburg im Breisgau, 1900.
- Neugebauer, Otto. *Astronomical Cuneiform Texts*. 3 volumes. London:1956; 2nd edition, New York: Springer, 1983. (Commonly abbreviated as *ACT*).
- Toomer, G. J. "Hipparchus and Babylonian Astronomy." In *A Scientific Humanist: Studies in Memory of Abraham Sachs*, ed. Erle Leichty, Maria deJ. Ellis, and Pamela Gerardi, pp. 353–362. Philadelphia: Occasional Publications of the Samuel Noah Kramer Fund 9, 1988.
- Watson, Rita; Horowitz, Wayne (2011). *Writing Science Before the Greeks: A Naturalistic Analysis of the Babylonian Astronomical Treatise MUL.APIN*. Leiden: Brill Academic Pub. ISBN 90-04-20230-7.

Babylonian astrology

Part of a series on

Ancient Mesopotamian religion

Related topics

- Ancient Near Eastern religions
- Sumerian religion
- Babylonian religion

- v
- t
- e[260]

Astrology

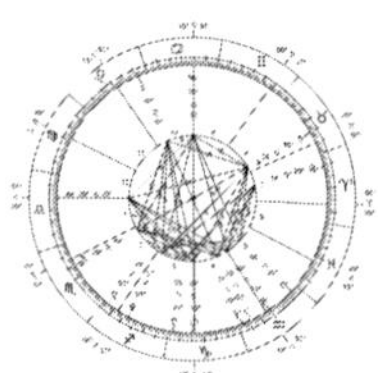

New millennium astrological chart

Background

- History of astrology
- Astrology and science
- Astrology and astronomy
- Traditions, types, and systems

Traditions

- Babylonian
- Hellenistic
- Islamic
- Western
- Hindu
- Chinese

Branches

- Natal
- Electional
- Horary

Astrology portal

- v
- t
- e[261]

In Babylon as well as in Assyria as a direct offshoot of Babylonian culture, astrology takes its place as one of the two chief means at the disposal of the priests (who were called *bare* or "inspectors") for ascertaining the will and intention of the gods, the other being through the inspection of the livers of sacrificial animals (see omen).

Early origins

Babylonian astrology was the first organized system of astrology, arising in the second millennium BC.[262] There is speculation that astrology of some form appeared in the Sumerian period in the 3rd millennium BC, but the isolated references to ancient celestial omens dated to this period are not considered sufficient evidence to demonstrate an integrated theory of astrology.[263] The history of scholarly celestial divination is therefore generally reported to begin with late Old Babylonian texts (c. 1800 BC), continuing through the Middle Babylonian and Middle Assyrian periods (c. 1200 BC).[264]

By the 16th century BC, the extensive employment of omen-based astrology can be evidenced in the compilation of a comprehensive reference work known as *Enuma Anu Enlil*. Its contents consisted of 70 cuneiform tablets comprising 7,000 celestial omens. Texts from this time also refer to an oral tradition – the origin and content of which can only be speculated upon.[265] At this time Babylonian astrology was solely mundane, and prior to the 7th century BC the practitioners' understanding of astronomy was fairly rudimentary. Because of their inability to accurately predict future celestial phenomena and planetary movement very far in advance, interpretations were done as the phenomena occurred or slightly before.Wikipedia:Citation needed By the 4th century, however, their mathematical methods had progressed enough to calculate future planetary positions with reasonable accuracy, at which point extensive ephemerides began to appear.[266]

Divinatory basis

The history of Babylonian astrology shows the development of astronomical knowledge within the context of divination. A collection of 32 tablets with inscribed liver models, dating from about 1875 BC, are the oldest known detailed texts of Babylonian divination, and these demonstrate the same interpretational format as that employed in celestial omen analysis.[267] Blemishes and marks found on the liver of the sacrificial animal were interpreted as symbolic signs which presented messages from the gods to the king.

The gods were also believed to present themselves in the celestial images of the planets or stars with whom they were associated. Evil celestial omens attached to any particular planet were therefore seen as indications of dissatisfaction or disturbance of the god that planet represented.[268] Such indications were met with attempts to appease the god and find manageable ways by which the god's expression could be realised without significant harm to the king and his nation. An astronomical report to the king Esarhaddon concerning a lunar eclipse of January 673 BC shows how the ritualistic use of substitute kings, or substitute events, combined an unquestioning belief in magic and omens with a purely mechanical view that the astrological event must have some kind of correlate within the natural world:

> *... In the beginning of the year a flood will come and break the dikes. When the Moon has made the eclipse, the king, my lord, should write to me. As a substitute for the king, I will cut through a dike, here in Babylonia, in the middle of the night. No one will know about it.*[269]

Ulla Koch-Westenholz, in her 1995 book *Mesopotamian Astrology*, argues that this ambivalence between a theistic and mechanic worldview defines the Babylonian concept of celestial divination as one which, despite its heavy reliance on magic, remains free of implications of targeted punishment with the purpose of revenge, and so "shares some of the defining traits of modern science: it is objective and value-free, it operates according to known rules, and its data are considered universally valid and can be looked up in written tabulations".[270] Koch-Westenholz also establishes the most important distinction between ancient Babylonian astrology and other divinatory disciplines as being that the former was originally exclusively mundane, being geographically oriented and specifically applied to countries cities and nations, and almost wholly concerned with the welfare of the state and the king as the governing head of the nation.[271]

Planets and gods

Of the planets five were recognized—Jupiter, Venus, Saturn, Mercury and Mars—to name them in the order in which they appear in the older cuneiform literature; in later texts Mercury and Saturn change places.

These five planets were identified with the gods of the Babylonian pantheon as follows:

- Jupiter with Marduk,
- Venus with the goddess Ishtar,
- Saturn with Ninurta (Ninib),
- Mercury with Nabu (Nebo),
- Mars with Nergal.

The movements of the Sun, Moon and five planets were regarded as representing the activity of the five gods in question, together with the moon-god Sin and the Sun-god Shamash, in preparing the occurrences on earth. If, therefore, one could correctly read and interpret the activity of these powers, one knew what the gods were aiming to bring about.

System of Interpretation

The Babylonian priests accordingly applied themselves to the task of perfecting a system of interpretation of the phenomena to be observed in the heavens, and it was natural that the system was extended from the moon, sun and five planets to the more prominent and recognizable fixed stars.

The interpretations themselves were based (as in the case of divination through the liver) chiefly on two factors:

- On the recollection or on written records of what in the past had taken place when the phenomenon or phenomena in question had been observed, and
- Association of ideas—involving sometimes merely a play upon words—in connection with the phenomenon or phenomena observed.

Figure 22: *Detail of the Ishtar Gate in Babylon*

Thus, if on a certain occasion, the rise of the new moon in a cloudy sky was followed by victory over an enemy or by abundant rain, the sign in question was thus proved to be a favourable one and its recurrence would thenceforth be regarded as a good omen, though the prognostication would not necessarily be limited to the one or the other of those occurrences, but might be extended to apply to other circumstances.

On the other hand, the appearance of the new moon earlier than was expected was regarded as unfavourable - prognosticating in one case defeat, in another death among cattle, in a third bad crops - not necessarily because these events actually took place after such a phenomenon, but by an application of the general principle resting upon association of ideas whereby anything premature would suggest an unfavourable occurrence.

In this way a mass of traditional interpretation of all kinds of observed phenomena was gathered, and once gathered became a guide to the priests for all times. However, not all of these ideas are still used in astrology as it is usually practiced today.

Limits of early knowledge

Astrology in its earliest stage was marked by three characteristic limitations:

General nature

The first limitation was that the movements and position of the heavenly bodies point to such occurrences as are of public import and affect the general welfare. The individual's interests are not in any way involved, and we must descend many centuries and pass beyond the confines of Babylonia and Assyria before we reach that phase which in medieval and modern astrology is almost exclusively dwelt upon—the individual horoscope.

In Babylonia and Assyria the cult centred largely and indeed almost exclusively in the public welfare and the person of the king, because upon his well-being and favour with the gods the fortunes of the country were dependent, in accordance with the ancient conception of kingship.[272]

Astronomical expertise

The second limitation was that the astronomical knowledge presupposed and accompanying early Babylonian astrology was, though essentially of an empirical character, limited and flawed. The theory of the ecliptic as representing the course of the Sun through the year, divided among twelve constellations with a measurement of 30° to each division, is of Babylonian origin, as has now been definitely proved; but it does not appear to have been perfected until after the fall of the Babylonian empire in 539 BC.

Similarly, the other accomplishments of Babylonian astronomers, such as their system or rather systems of moon calculations and the drawing up of planetary tablets, belong to this late period, so that the golden age of Babylonian astronomy belongs not to the remote past, as was until recently supposed, but to the Seleucid period; i.e. after the advent of the Greeks in the Euphrates Valley.

From certain expressions used in astrological texts that are earlier than the 7th century BC it would appear, indeed, that the beginnings at least of the calculation of sun and moon eclipses belong to the earlier period, but here, too, the chief work accomplished was after 400 BC, and the defectiveness of early Babylonian astronomy may be gathered from the fact that as late as the 6th century BC an error of almost an entire month was made by the Babylonian astronomers in the attempt to determine through calculation the beginning of a certain year.

In a general way, the reign of law and order in the movements of the heavenly bodies was recognized, and indeed must have exercised an influence at an early period in leading to the rise of a methodical divination that was certainly of a much higher order than the examination of an animal's liver.

However, the importance that was laid upon the endless variations in the form of the phenomena and the equally numerous apparent deviations from what were regarded as normal conditions, prevented for a long time the rise of any serious study of astronomy beyond what was needed for the purely practical purposes that the priests as "inspectors" of the heavens (as they were also the "inspectors" of the sacrificial livers) had in mind.

Constellations

The third limitation was that there is little evidence that the signs of the zodiac that we now recognise, were used in Babylonian astronomy prior to 700 BC. However, probably from as early as the days of Hammurabi c. 2000 BC, Babylonian astrologers did develop the idea of constellations by depicting prominent groups of stars with outlines of images derived from their mythology and religion.

Ashurbanipal

Ashurbanipal was a king of Assyria who ruled in the seventh century BC from 668 to 625 BC.[273] He was famous for assembling a great library of cuneiform tablets in Nineveh on the subjects of astrology, history, mythology, and science. Some of Assurbanipal's astrologers, such as Rammanu-sumausar and Nabu-musisi, became so adept at deducing omens from daily movements of the planets that a system of making periodical reports to the king came into being. Thus, Assurbanipal received swift messengers detailing "all occurrences in heaven and earth" throughout his kingdom and the results of his astrologer's examinations of them. He then used this information as a political weapon, and for the practical day-to-day running of his kingdom. After his death Nineveh fell to the Medes and the Chaldean Babylonians, and Assurbanipal's library was destroyed or dispersed.

Sources

- Baigent, Michael, 1994. *From the Omens of Babylon: Astrology and Ancient Mesopotamia*. Arkana. ISBN 0-14-019480-0.
- Holden, James Herschel, 1996. *A History of Horoscopic Astrology*. AFA. ISBN 978-0-86690-463-6.
- Koch-Westenholz, Ulla, 1995. *Mesopotamian astrology*. Volume 19 of CNI publications. Museum Tusculanum Press. ISBN 978-87-7289-287-0.

- Neugebauer, Otto, 1969 (1st edition: 1957). *The Exact Sciences in Antiquity* (2nd ed.) Dover Publications. ISBN 978-0-486-22332-2.
- Rochberg, Francesa, 1998. *Babylonian Horoscopes*. American Philosophical Society. ISBN 0-87169-881-1.
- Verderame, Lorenzo, "The Primeval Zodiac: Its Social, Religious, and Mythological Background", in J.A. Rubiño-Martín et al., *Cosmology Across Cultures*, ASP Conference Series 409, San Francisco, 2009, 151-156.

Babylonian calendar

The **Babylonian calendar** was a lunisolar calendar with years consisting of 12 lunar months, each beginning when a new crescent moon was first sighted low on the western horizon at sunset, plus an intercalary month inserted as needed by decree. The calendar is based on a Sumerian (Ur III) predecessor preserved in the **Umma calendar** of Shulgi (c. 21st century BC).

Months

The year begins in spring, and is divided into *reš šatti* "beginning", *mišil šatti* "middle", and *kīt šatti* "end of the year". The word for "month" was *arḫu* (construct state *araḫ*). That the calendar originates in Babylonian, and not in later Assyrian times is shown by the fact that the chief deity of the Assyrians is assigned the surplus intercalary month. During the 6th century BC Babylonian captivity of the Hebrews, the Babylonian month names were adopted into the Hebrew calendar. The Assyrian calendar used in Iraq and the Levant also uses many of the same names for its months, such as Iyyar, Tammuz, Ab, Elul, Tishri, and Adar.

Season		Month name	Presiding deities	Zodiac sign	Hebrew equivalent	Gregorian equivalent
Reš Šatti	1	**Araḫ Nisānu** 'Month of the Sanctuary'	Anu and Bel	KU (Aries)	Nisan	March/April
	2	**Araḫ Āru** 'Month of the Bull'	Ea	(Taurus)	Iyar	April/May
	3	**Araḫ Simanu**	Sīn	BI(KAŠ) (Gemini)	Sivan	May/June
	4	**Araḫ Dumuzu** 'Month of Tammuz'	Tammuz	(Cancer)	Tammuz	June/July

Mišil Šatti	5	**Araḫ Abu**		āru (Leo)	Av	July/August
	6	**Araḫ Ulūlu**	Ishtar	(Virgo)	Elul	August/-September
	7	**Araḫ Tišritum** 'Month of Beginning' (i.e. the start of the second half-year)	Shamash	(Libra)	Tishrei	September/-October
	8	**Araḫ Samna** 'Month of Laying Foundations'	Marduk	(Scorpio)	Cheshvan	October/-November
Kīt Šatti	9	**Araḫ Kislimu**	Nergal	(Sagit-tarius)	Kislev	November/-December
	10	**Araḫ Ṭebētum** 'Month of the Forthcoming of Water'	Papsukkal	saḫ 'ibex' (Capri-corn?)	Tevet	December/-January
	11	**Araḫ Šabaṭu**		qā (Aquarius?)	Shevat	January/-February
	12	**Araḫ Addaru ~ Araḫ Adār** 'Month of Adar'	Erra	(Pisces)	Adar	February/-March
Inter-calary	13	**Araḫ Makaruša Addari**	Assur	In year 17 of 19-year cycle, the inter-calary month was named *Araḫ Ulūlu.*		

|+ Babylonian calendar

Until the 5th century BC, the calendar was fully observational, but beginning about 499 BC the months began to be regulated by a lunisolar cycle of 19 years equaling 235 months. Although usually called the Metonic cycle after Meton of Athens (432 BC), Meton probably learned of the cycle from the Babylonians. After no more than three isolated exceptions, by 380 BC the months of the calendar were regulated by the cycle without exception. In the cycle of 19 years, the month **Adaru 2** was intercalated, except in the year that was number 17 in the cycle, when the month **Ululu 2** was inserted. During this period, the first day of each month (beginning at sunset) continued to be the day when a new crescent moon was first sighted—the calendar never used a specified number of days in any month.

Days

Counting from the new moon, the Babylonians celebrated every seventh day as a "holy-day", also called an "evil-day" (meaning "unsuitable" for prohibited activities) .Wikipedia:Citation needed On these days officials were prohibited from various activities and common men were forbidden to "make a wish", Wikipedia:Citation needed and at least the 28th was known as a "rest-day" .Wikipedia:Citation needed On each of them, offerings were made to a different god and goddess, apparently at nightfall to avoid the prohibitions: Marduk and Ishtar on the 7th, Ninlil and Nergal on the 14th, Sin and Shamash on the 21st, and Enki and Mah on the 28th. Tablets from the sixth-century BC reigns of Cyrus the Great and Cambyses II indicate these dates were sometimes approximate. The lunation of 29 or 30 days basically contained three seven-day weeks, and a final week of eight or nine days inclusive, breaking the continuous seven-day cycle.

Among other theories of *Shabbat* origin, the *Universal Jewish Encyclopedia* of Isaac Landman advanced a theory of Assyriologists like Friedrich Delitzsch that *Shabbat* originally arose from the lunar cycle, containing four weeks ending in Sabbath, plus one or two additional unreckoned days per month. The difficulties of this theory include reconciling the differences between an unbroken week and a lunar week, and explaining the absence of texts naming the lunar week as *Shabbat* in any language.

The Babylonians additionally celebrated the 19th as a special "evil day", the "day of anger", because it was roughly the 49th day of the (preceding) month, completing a "week of weeks".Wikipedia:Citation needed Sacrifices were offered to Ninurta and the day dedicated to Gula, and it may be supposed that prohibitions were strengthened.Wikipedia:Citation needed

Further, reconstruction of a broken tabletWikipedia:Citation needed seems to define the rarely attested *Sapattu*^{m} or *Sabattu*^{m} as the full moon. This word is cognate or merged with Hebrew *Shabbat*, but is monthly rather than weekly; it is regarded as a form of Sumerian *sa-bat* ("mid-rest"), attested in Akkadian as *um nuh libbi* ("day of mid-repose"). According to Marcello Craveri, Sabbath "was almost certainly derived from the Babylonian *Shabattu*, the festival of the full moon, but, all trace of any such origin having been lost, the Hebrews ascribed it to Biblical legend." This conclusion is a contextual restoration of the damaged *Enûma Eliš* creation account, which is read as: "[Sa]bbath shalt thou then encounter, mid[month]ly."

References

Bibliography

- Parker, Richard Anthony and Waldo H. Dubberstein. *Babylonian Chronology 626 BC.–AD. 75*. Providence, RI: Brown University Press, 1956.
- Structure of the Babylonian calendar[274]
- W. Muss-Arnolt, *The Names of the Assyro-Babylonian Months and Their Regents*, Journal of Biblical Literature (1892).
- Sacha Stern, "The Babylonian Calendar at Elephantine" in *Zeitschrift für Papyrologie und Epigraphik* 130 (2000) 159–171 (PDF document[275], 94KB)

- Fales, Frederick Mario, "A List of Umma Month Names", Revue d'assyriologie et d'archéologie orientale, 76 (1982), 70–71.
- Gomi, Tohru, "On the Position of the Month iti-ezem-dAmar-dSin in the Neo-Sumerian Umma Calendar", Zeitschrift für Assyriologie und Vorderasiatische Archäologie, 75 (1985), 4–6.
- Pomponio, Francesco, "The Reichskalender of Ur III in the Umma Texts", Zeitschrift für Assyriologie und Vorderasiastische Archäologie, 79 (1989), 10–13.
- Verderame, Lorenzo, "Le calendrier et le compte du temps dans la pensée mythique suméro-akkadienne", De Kêmi à Birit Nâri, Revue Internationale de l'Orient Ancien, 3 (2008), 121–134.
- Steele, John M., ed., "Calendars and Years: Astronomy and Time in the Ancient Near East", Oxford: Oxbow, 2007.

External links

- The Babylonian Ritual Calendar[276]
- The Mul.Apin Tablets[277]
- The Babylonian Calendar (with a date converter based on Parker & Dubberstein (1971))[278]

Indian astronomy

Indian astronomy has a long history stretching from pre-historic to modern times. Some of the earliest roots of Indian astronomy can be dated to the period of Indus Valley Civilization or earlier. Astronomy later developed as a discipline of Vedanga or one of the "auxiliary disciplines" associated with the study of the Vedas,[279] dating 1500 BCE or older.[280] The oldest known text is the *Vedanga Jyotisha*, dated to 1400–1200 BCE (with the extant form possibly from 700–600 BCE).

Indian astronomy was influenced by Greek astronomy beginning in the 4th century BCE[281,282,283] and through the early centuries of the Common Era, for example by the Yavanajataka and the *Romaka Siddhanta*, a Sanskrit translation of a Greek text disseminated from the 2nd century.[284]

Indian astronomy flowered in the 5th-6th century, with Aryabhata, whose *Aryabhatiya* represented the pinnacle of astronomical knowledge at the time. Later the Indian astronomy significantly influenced Muslim astronomy, Chinese astronomy, European astronomy,[285] and others. Other astronomers of the classical era who further elaborated on Aryabhata's work include Brahmagupta, Varahamihira and Lalla.

An identifiable native Indian astronomical tradition remained active throughout the medieval period and into the 16th or 17th century, especially within the Kerala school of astronomy and mathematics.

History

Some of the earliest forms of astronomy can be dated to the period of Indus Valley Civilization, or earlier. Some cosmological concepts are present in the Vedas, as are notions of the movement of heavenly bodies and the course of the year. As in other traditions, there is a close association of astronomy and religion during the early history of the science, astronomical observation being necessitated by spatial and temporal requirements of correct performance of religious ritual. Thus, the *Shulba Sutras*, texts dedicated to altar construction, discusses advanced mathematics and basic astronomy. *Vedanga Jyotisha* is another of the earliest known Indian texts on astronomy, it includes the details about the sun, moon, nakshatras, lunisolar calendar.[286]

Greek astronomical ideas began to enter India in the 4th century BCE following the conquests of Alexander the Great.[284] By the early centuries of the Common Era, Indo-Greek influence on the astronomical tradition is visible, with texts such as the Yavanajataka and *Romaka Siddhanta*. Later astronomers mention the existence of various siddhantas during this period, among them a text known as the *Surya Siddhanta*. These were not fixed texts but rather an oral tradition of knowledge, and their content is not extant. The text today known as *Surya Siddhanta* dates to the Gupta period and was received by Aryabhata.

The classical era of Indian astronomy begins in the late Gupta era, in the 5th to 6th centuries. The *Pañcasiddhāntikā* by Varāhamihira (505 CE) approximates the method for determination of the meridian direction from any three positions of the shadow using a gnomon. By the time of Aryabhata the motion of planets was treated to be elliptical rather than circular.[287] Other topics included definitions of different units of time, eccentric models of planetary motion, epicyclic models of planetary motion, and planetary longitude corrections for various terrestrial locations.

Calendars

The divisions of the year were on the basis of religious rites and seasons (*Rtu*).[288] The duration from mid March—Mid May was taken to be spring (*vasanta*), mid May—mid July: summer ("grishma"), mid July—mid September: rains (*varsha*), mid September—mid November: autumn, mid November—mid January: winter, mid January—mid March: *dew* (*śiśira*).

Figure 23: *A page from the Hindu calendar 1871–72.*

In the *Vedānga Jyotiṣa*, the year begins with the winter solstice.[289] Hindu calendars have several eras:

- The Hindu calendar, counting from the start of the Kali Yuga, has its epoch on 18 February 3102 BCE Julian (23 January 3102 BCE Gregorian).
- The Vikrama Samvat calendar, introduced about the 12th century, counts from 56–57 BCE.
- The "Saka Era", used in some Hindu calendars and in the Indian national calendar, has its epoch near the vernal equinox of year 78.
- The Saptarshi calendar traditionally has its epoch at 3076 BCE.[290]

J.A.B. van Buitenen (2008) reports on the calendars in India:

> *The oldest system, in many respects the basis of the classical one, is known from texts of about 1000 BCE. It divides an approximate solar year of 360 days into 12 lunar months of 27 (according to the early Vedic text Taittirīya Saṃhitā 4.4.10.1–3) or 28 (according to the Atharvaveda, the fourth of the Vedas, 19.7.1.) days. The resulting discrepancy was resolved by the intercalation of a leap month every 60 months. Time was reckoned by the position marked off in constellations on the ecliptic in which the Moon rises daily in the course of one lunation (the period from New Moon to New Moon) and the Sun rises monthly in the course of one year. These*

constellations (nakṣatra) each measure an arc of 13° 20′ of the ecliptic circle. The positions of the Moon were directly observable, and those of the Sun inferred from the Moon's position at Full Moon, when the Sun is on the opposite side of the Moon. The position of the Sun at midnight was calculated from the nakṣatra that culminated on the meridian at that time, the Sun then being in opposition to that nakṣatra.

Astronomers

Name	Year	Contributions
Lagadha	1st millennium BCE	The earliest astronomical text—named *Vedānga Jyotiṣa* details several astronomical attributes generally applied for timing social and religious events. The *Vedānga Jyotiṣa* also details astronomical calculations, calendrical studies, and establishes rules for empirical observation.[291] Since the texts written by 1200 BCE were largely religious compositions the *Vedānga Jyotiṣa* has connections with Indian astrology and details several important aspects of the time and seasons, including lunar months, solar months, and their adjustment by a lunar leap month of *Adhimāsa*.[292] *Ritus* are also described as ((yugams)). Tripathi (2008) holds that ' Twenty-seven constellations, eclipses, seven planets, and twelve signs of the zodiac were also known at that time.'
Aryabhata	476–550 CE	Aryabhata was the author of the *Āryabhatīya* and the *Aryabhatasiddhanta*, which, according to Hayashi (2008): 'circulated mainly in the northwest of India and, through the Sāsānian dynasty (224–651) of Iran, had a profound influence on the development of Islamic astronomy. Its contents are preserved to some extent in the works of Varahamihira (flourished c. 550), Bhaskara I (flourished c. 629), Brahmagupta (598–c. 665), and others. It is one of the earliest astronomical works to assign the start of each day to midnight.' Aryabhata explicitly mentioned that the earth rotates about its axis, thereby causing what appears to be an apparent westward motion of the stars. In his book, Aryabhatiya, he suggested that the Earth was sphere, containing a circumference of 24,835 miles (39,967 km).[293] Aryabhata also mentioned that reflected sunlight is the cause behind the shining of the moon. Aryabhata's followers were particularly strong in South India, where his principles of the diurnal rotation of the earth, among others, were followed and a number of secondary works were based on them.
Brahmagupta	598–668 CE	*Brahmasphuta-siddhanta* (Correctly Established Doctrine of Brahma, 628 CE) dealt with both Indian mathematics and astronomy. Hayashi (2008) writes: 'It was translated into Arabic in Baghdad about 771 and had a major impact on Islamic mathematics and astronomy.'[294] In *Khandakhadyaka* (A Piece Eatable, 665 CE) Brahmagupta reinforced Aryabhata's idea of another day beginning at midnight. Bahmagupta also calculated the instantaneous motion of a planet, gave correct equations for parallax, and some information related to the computation of eclipses. His works introduced Indian concept of mathematics based astronomy into the Arab world. He also theorized that all bodies with mass are attracted to the earth.[295]
Varāhamihira	505 CE	Varāhamihira was an astronomer and mathematician who studied and Indian astronomy as well as the many principles of Greek, Egyptian, and Roman astronomical sciences. His *Pañcasiddhāntikā* is a treatise and compendium drawing from several knowledge systems.[296]

Bhāskara I	629 CE	Authored the astronomical works *Mahabhaskariya* (Great Book of Bhaskara), *Laghubhaskariya* (Small Book of Bhaskara), and the *Aryabhatiyabhashya* (629 CE)—a commentary on the *Āryabhatīya* written by Aryabhata. Hayashi (2008) writes 'Planetary longitudes, heliacal rising and setting of the planets, conjunctions among the planets and stars, solar and lunar eclipses, and the phases of the Moon are among the topics Bhaskara discusses in his astronomical treatises.'[294] Baskara I's works were followed by Vateśvara (880 CE), who in his eight chapter *Vateśvarasiddhānta* devised methods for determining the parallax in longitude directly, the motion of the equinoxes and the solstices, and the quadrant of the sun at any given time.
Lalla	8th century CE	Author of the *Śisyadhīvrddhida* (Treatise Which Expands the Intellect of Students), which corrects several assumptions of Āryabhata. The *Śisyadhīvrddhida* of Lalla itself is divided into two parts:*Grahādhyāya* and *Golādhyāya*. *Grahādhyāya* (Chapter I-XIII) deals with planetary calculations, determination of the mean and true planets, three problems pertaining to diurnal motion of Earth, eclipses, rising and setting of the planets, the various cusps of the moon, planetary and astral conjunctions, and complementary situations of the sun and the moon.[297] The second part—titled *Golādhyāya* (chapter XIV–XXII)—deals with graphical representation of planetary motion, astronomical instruments, spherics, and emphasizes on corrections and rejection of flawed principles. Lalla shows influence of Āryabhata, Brahmagupta, and Bhāskara I. His works were followed by later astronomers Śrīpati, Vateśvara, and Bhāskara II. Lalla also authored the *Siddhāntatilaka*.
Bhāskara II	1114 CE	Authored *Siddhāntaśiromaṇi* (Head Jewel of Accuracy) and *Karaṇakutūhala* (Calculation of Astronomical Wonders) and reported on his observations of planetary positions, conjunctions, eclipses, cosmography, geography, mathematics, and astronomical equipment used in his research at the observatory in Ujjain, which he headed.[294]
Śrīpati	1045 CE	Śrīpati was an astronomer and mathematician who followed the Brhmagupta school and authored the *Siddhāntaśekhara* (The Crest of Established Doctrines) in 20 chapters, thereby introducing several new concepts, including moon's second inequality.[298]
Mahendra Suri	14th century CE	Mahendra Suri authored the *Yantra-rāja* (The King of Instruments, written in 1370 CE)—a Sanskrit work on the astrolabe, itself introduced in India during the reign of the 14th century Tughlaq dynasty ruler Firuz Shah Tughluq (1351–1388 CE). Suri seems to have been a Jain astronomer in the service of Firuz Shah Tughluq. The 182 verse *Yantra-rāja* mentions the astrolabe from the first chapter onwards, and also presents a fundamental formula along with a numerical table for drawing an astrolabe although the proof itself has not been detailed. Longitudes of 32 stars as well as their latitudes have also been mentioned. Mahendra Suri also explained the Gnomon, equatorial co-ordinates, and elliptical co-ordinates. The works of Mahendra Suri may have influenced later astronomers like Padmanābha (1423 CE)—author of the *Yantra-rāja-adhikāra*, the first chapter of his *Yantra-kirnāvali*.[299]

Nilakan-than Somayaji	1444–1544 CE	In 1500, Nilakanthan Somayaji of the Kerala school of astronomy and mathematics, in his *Tantrasangraha*, revised Aryabhata's model for the planets Mercury and Venus. His equation of the centre for these planets remained the most accurate until the time of Johannes Kepler in the 17th century.[300] Nilakanthan Somayaji, in his *Aryabhatiyabhasya*, a commentary on Aryabhata's *Aryabhatiya*, developed his own computational system for a partially heliocentric planetary model, in which Mercury, Venus, Mars, Jupiter and Saturn orbit the Sun, which in turn orbits the Earth, similar to the Tychonic system later proposed by Tycho Brahe in the late 16th century. Nilakantha's system, however, was mathematically more efficient than the Tychonic system, due to correctly taking into account the equation of the centre and latitudinal motion of Mercury and Venus. Most astronomers of the Kerala school of astronomy and mathematics who followed him accepted his planetary model.[301] He also authored a treatise titled Jyotirmimamsa stressing the necessity and importance of astronomical observations to obtain correct parameters for computations.
Acyuta Pisārati	1550–1621 CE	*Sphutanirnaya* (Determination of True Planets) details an elliptical correction to existing notions. *Sphutanirnaya* was later expanded to *Rāśigolasphutānīti* (True Longitude Computation of the Sphere of the Zodiac).[279] Another work, *Karanottama* deals with eclipses, complementary relationship between the sun and the moon, and 'the derivation of the mean and true planets'. In *Uparāgakriyākrama* (Method of Computing Eclipses), Acyuta Pisārati suggests improvements in methods of calculation of eclipses.

Instruments used

Among the devices used for astronomy was gnomon, known as *Sanku*, in which the shadow of a vertical rod is applied on a horizontal plane in order to ascertain the cardinal directions, the latitude of the point of observation, and the time of observation.[302] This device finds mention in the works of Varāhamihira, Āryabhata, Bhāskara, Brahmagupta, among others.[303] The Cross-staff, known as *Yasti-yantra*, was used by the time of Bhaskara II (1114–1185 CE). This device could vary from a simple stick to V-shaped staffs designed specifically for determining angles with the help of a calibrated scale. The clepsydra (*Ghatī-yantra*) was used in India for astronomical purposes until recent times. Ōhashi (2008) notes that: "Several astronomers also described water-driven instruments such as the model of fighting sheep."

The armillary sphere was used for observation in India since early times, and finds mention in the works of Āryabhata (476 CE).[304] The *Goladīpikā*—a detailed treatise dealing with globes and the armillary sphere was composed between 1380–1460 CE by Parameśvara. On the subject of the usage of the armillary sphere in India, Ōhashi (2008) writes: "The Indian armillary sphere (*gola-yantra*) was based on equatorial coordinates, unlike the Greek armillary sphere, which was based on ecliptical coordinates, although the Indian armillary sphere also had an ecliptical hoop. Probably, the celestial coordinates of the junction stars of the lunar mansions were determined by the armillary sphere since the seventh century or so. There was also a celestial globe rotated by flowing water."

Figure 24: *Sawai Jai Singh (1688–1743 CE) initiated the construction of several observatories. Shown here is the Jantar Mantar (Jaipur) observatory.*

Figure 25: *Yantra Mandir (completed by 1743 CE), Delhi.*

Figure 26: *Astronomical instrument with graduated scale and notation in Hindu-Arabic numerals.*

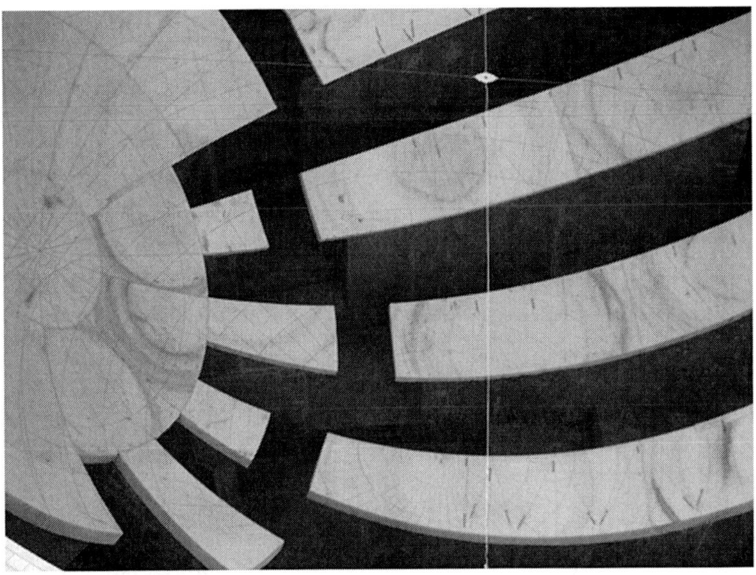

Figure 27: *Detail of an instrument in the Jaipur observatory.*

An instrument invented by the mathematician and astronomer Bhaskara II (1114–1185 CE) consisted of a rectangular board with a pin and an index arm. This device—called the *Phalaka-yantra*—was used to determine time from the sun's altitude. The *Kapālayantra* was an equatorial sundial instrument used to determine the sun's azimuth. *Kartarī-yantra* combined two semicircular board instruments to give rise to a 'scissors instrument'. Introduced from the Islamic world and first finding mention in the works of Mahendra Sūri—the court astronomer of Firuz Shah Tughluq (1309–1388 CE)—the astrolabe was further mentioned by Padmanābha (1423 CE) and Rāmacandra (1428 CE) as its use grew in India.

Invented by *Padmanābha*, a nocturnal polar rotation instrument consisted of a rectangular board with a slit and a set of pointers with concentric graduated circles. Time and other astronomical quantities could be calculated by adjusting the slit to the directions of α and β Ursa Minor. Ōhashi (2008) further explains that: "Its backside was made as a quadrant with a plumb and an index arm. Thirty parallel lines were drawn inside the quadrant, and trigonometrical calculations were done graphically. After determining the sun's altitude with the help of the plumb, time was calculated graphically with the help of the index arm."

Ōhashi (2008) reports on the observatories constructed by Jai Singh II of Amber:

> *The Mahārāja of Jaipur, Sawai Jai Singh (1688–1743 CE), constructed five astronomical observatories at the beginning of the eighteenth century. The observatory in Mathura is not extant, but those in Delhi, Jaipur, Ujjain, and Banaras are. There are several huge instruments based on Hindu and Islamic astronomy. For example, the samrāt.-yantra (emperor instrument) is a huge sundial which consists of a triangular gnomon wall and a pair of quadrants toward the east and west of the gnomon wall. Time has been graduated on the quadrants.*

The seamless celestial globe invented in Mughal India, specifically Lahore and Kashmir, is considered to be one of the most impressive astronomical instruments and remarkable feats in metallurgy and engineering. All globes before and after this were seamed, and in the 20th century, it was believed by metallurgists to be technically impossible to create a metal globe without any seams, even with modern technology. It was in the 1980s, however, that Emilie Savage-Smith discovered several celestial globes without any seams in Lahore and Kashmir. The earliest was invented in Kashmir by Ali Kashmiri ibn Luqman in 1589–90 CE during Akbar the Great's reign; another was produced in 1659–60 CE by Muhammad Salih Tahtawi with Arabic and Sanskrit inscriptions; and the last was produced in Lahore by a Hindu metallurgist Lala Balhumal Lahuri in 1842 during Jagatjit Singh Bahadur's reign. 21 such globes

Figure 28: *Greek equatorial sun dial, Ai-Khanoum, Afghanistan 3rd–2nd century BCE.*

were produced, and these remain the only examples of seamless metal globes. These Mughal metallurgists developed the method of lost-wax casting in order to produce these globes.[305]

Global discourse

Indian and Greek astronomy

The earliest known Indian astronomical work (though it is restricted to calendrical discussions) is the *Vedanga Jyotisha* of Lagadha, which is dated to 1400–1200 BCE (with the extant form possibly from 700–600 BCE). According to Pingree, there are a number of Indian astronomical texts that are dated to the sixth century CE or later with a high degree of certainty. There is substantial similarity between these and pre-Ptolomaic Greek astronomy. Pingree believes that these similarities suggest a Greek origin for certain aspects of Indian astronomy. This has been contested by Bartel Leendert van der Waerden who maintains the originality and independent development of Indian astronomy.[306]

With the rise of Greek culture in the east, Hellenistic astronomy filtered eastwards to India, where it profoundly influenced the local astronomical tradition. For example, Hellenistic astronomy is known to have been practiced near India in the Greco-Bactrian city of Ai-Khanoum from the 3rd century BCE. Various sun-dials, including an equatorial sundial adjusted to the latitude of Ujjain have been found in archaeological excavations there.[307] Numerous interactions with the Mauryan Empire, and the later expansion of the Indo-Greeks into India suggest that transmission of Greek astronomical ideas to India occurred during this period.[308] The Greek concept of a spherical earth surrounded by the spheres of planets, further influenced the astronomers like Varahamihira and Brahmagupta.[309]

Several Greco-Roman astrological treatises are also known to have been exported to India during the first few centuries of our era. The *Yavanajataka* was a Sanskrit text of the 3rd century CE on Greek horoscopy and mathematical astronomy. Rudradaman's capital at Ujjain "became the Greenwich of Indian astronomers and the Arin of the Arabic and Latin astronomical treatises; for it was he and his successors who encouraged the introduction of Greek horoscopy and astronomy into India."[310]

Later in the 6th century, the *Romaka Siddhanta* ("Doctrine of the Romans"), and the *Paulisa Siddhanta* ("Doctrine of Paul") were considered as two of the five main astrological treatises, which were compiled by Varāhamihira in his *Pañca-siddhāntikā* ("Five Treatises").[311] Varāhamihira goes on to state that "The Greeks, indeed, are foreigners, but with them this science (astronomy) is in a flourishing state." Another Indian text, the *Gargi-Samḥita*, also similarly compliments the Yavanas (Greeks) noting that the Yavanas though barbarians must be respected as seers for their introduction of astronomy in India.

Indian and Chinese astronomy

Indian astronomy reached China with the expansion of Buddhism during the Later Han (25–220 CE). Further translation of Indian works on astronomy was completed in China by the Three Kingdoms era (220–265 CE).[312] However, the most detailed incorporation of Indian astronomy occurred only during the Tang Dynasty (618–907 CE) when a number of Chinese scholars—such as Yi Xing— were versed both in Indian and Chinese astronomy. A system of Indian astronomy was recorded in China as *Jiuzhi-li* (718 CE), the author of which was an Indian by the name of Qutan Xida—a translation of Devanagari Gotama Siddha—the director of the Tang dynasty's national astronomical observatory.

Fragments of texts during this period indicate that Arabs adopted the sine function (inherited from Indian mathematics) instead of the chords of arc used in Hellenistic mathematics.[313] Another Indian influence was an approximate formula used for timekeeping by Muslim astronomers.[314] Through Islamic astronomy, Indian astronomy had an influence on European astronomy via Arabic translations. During the Latin translations of the 12th century, Muhammad al-Fazari's *Great Sindhind* (based on the *Surya Siddhanta* and the works of Brahmagupta), was translated into Latin in 1126 and was influential at the time.[315]

Indian and Islamic astronomy

In the 17th century, the Mughal Empire saw a synthesis between Islamic and Hindu astronomy, where Islamic observational instruments were combined with Hindu computational techniques. While there appears to have been little concern for planetary theory, Muslim and Hindu astronomers in India continued to make advances in observational astronomy and produced nearly a hundred *Zij* treatises. Humayun built a personal observatory near Delhi, while Jahangir and Shah Jahan were also intending to build observatories but were unable to do so. After the decline of the Mughal Empire, it was a Hindu king, Jai Singh II of Amber, who attempted to revive both the Islamic and Hindu traditions of astronomy which were stagnating in his time. In the early 18th century, he built several large observatories called Yantra Mandirs in order to rival Ulugh Beg's Samarkand observatory and in order to improve on the earlier Hindu computations in the *Siddhantas* and Islamic observations in *Zij-i-Sultani*. The instruments he used were influenced by Islamic astronomy, while the computational techniques were derived from Hindu astronomy.[316,317]

Indian astronomy and Europe

Some scholars have suggested that knowledge of the results of the Kerala school of astronomy and mathematics may have been transmitted to Europe through the trade route from Kerala by traders and Jesuit missionaries. Kerala was in continuous contact with China, Arabia and Europe. The existence of circumstantial evidence[318] such as communication routes and a suitable chronology certainly make such a transmission a possibility. However, there is no direct evidence by way of relevant manuscripts that such a transmission took place.[319]

In the early 18th century, Jai Singh II of Amber invited European Jesuit astronomers to one of his Yantra Mandir observatories, who had bought back the astronomical tables compiled by Philippe de La Hire in 1702. After examining La Hire's work, Jai Singh concluded that the observational techniques and instruments used in European astronomy were inferior to those used in India at the time - it is uncertain whether he was aware of the Copernican Revolution via the Jesuits.[320] He did, however, employ the use of telescopes. In his *Zij-i Muhammad Shahi*, he states: "telescopes were constructed in my kingdom and using them a number of observations were carried out".

Following the arrival of the British East India Company in the 18th century, the Hindu and Islamic traditions were slowly displaced by European astronomy, though there were attempts at harmonising these traditions. The Indian scholar Mir Muhammad Hussain had travelled to England in 1774 to study Western science and, on his return to India in 1777, he wrote a Persian treatise on astronomy. He wrote about the heliocentric model, and argued that there exists an infinite number of universes (*awalim*), each with their own planets and stars, and that this demonstrates the omnipotence of God, who is not confined to a single universe. Hussain's idea of a universe resembles the modern concept of a galaxy, thus his view corresponds to the modern view that the universe consists of billions of galaxies, each one consisting of billions of stars. The last known *Zij* treatise was the *Zij-i Bahadurkhani*, written in 1838 by the Indian astronomer Ghulam Hussain Jaunpuri (1760–1862) and printed in 1855, dedicated to Bahadur Khan. The treatise incorporated the heliocentric system into the *Zij* tradition.

Further reading

- Brennand, William (1896), *Hindu Astronomy*[321], Chas.Straker & Sons, London
- Maunder, E. Walter (1899), *The Indian Eclipse 1898*[322], Hazell Watson and Viney Ltd., London

References

- Abraham, G. (2008), "Gnomon in India", *Encyclopaedia of the History of Science, Technology, and Medicine in Non-Western Cultures (2nd edition)* edited by Helaine Selin, pp. 1035–1037, Springer, ISBN 978-1-4020-4559-2.
- Almeida, D. F.; et al. (2001). "Keralese Mathematics: Its Possible Transmission to Europe and the Consequential Educational Implications". *Journal of Natural Geometry*. **20**: 77–104.
- Baber, Zaheer (1996), *The Science of Empire: Scientific Knowledge, Civilization, and Colonial Rule in India*, State University of New York Press, ISBN 0-7914-2919-9.
- Dallal, Ahmad (1999), "Science, Medicine and Technology", *The Oxford History of Islam* edited by John Esposito, Oxford University Press.
- Hayashi, Takao (2008), *Aryabhata I*, Encyclopædia Britannica.
- Hayashi, Takao (2008), *Bhaskara I*, Encyclopædia Britannica.
- Hayashi, Takao (2008), *Brahmagupta*, Encyclopædia Britannica.

- Hayashi, Takao (2008), *Shripati*, Encyclopædia Britannica.
- J.A.B. van Buitenen (2008), *calendar*, Encyclopædia Britannica.
- Joseph, George G. (2000), *The Crest of the Peacock: Non-European Roots of Mathematics*, Penguin Books, ISBN 0-691-00659-8.
- King, David A (2002). "A Vetustissimus Arabic Text on the Quadrans Vetus". *Journal for the History of Astronomy*. **33**: 237–255. Bibcode: 2002JHA....33..237K[323].
- Klostermaier, Klaus K. (2003), "Hinduism, History of Science and Religion", *Encyclopedia of Science and Religion* edited by J. Wentzel Vrede van Huyssteen, pp. 405–410, Macmillan Reference USA, ISBN 0-02-865704-7.
- Raju, C. K. (2001). "Computers, Mathematics Education, and the Alternative Epistemology of the Calculus in the Yuktibhasa". *Philosophy East and West*. **51** (3): 325–362. doi: 10.1353/pew.2001.0045[324].
- Ramasubramanian (1994). "Modification of the earlier Indian planetary theory by the Kerala astronomers (c. 1500 CE) and the implied heliocentric picture of planetary motion". *Current Science*. **66**: 784–790.
- Sarma, K.V. (2008), "Acyuta Pisarati", *Encyclopaedia of the History of Science, Technology, and Medicine in Non-Western Cultures (2nd edition)* edited by Helaine Selin, p. 19, Springer, ISBN 978-1-4020-4559-2.
- Sarma, K.V. (2008), "Armillary Spheres in India", *Encyclopaedia of the History of Science, Technology, and Medicine in Non-Western Cultures (2nd edition)* edited by Helaine Selin, p. 243, Springer, ISBN 978-1-4020-4559-2.
- Sarma, K.V. (2008), "Astronomy in India", *Encyclopaedia of the History of Science, Technology, and Medicine in Non-Western Cultures (2nd edition)* edited by Helaine Selin, pp. 317–321, Springer, ISBN 978-1-4020-4559-2.
- Sarma, K.V. (2008), "Lalla", *Encyclopaedia of the History of Science, Technology, and Medicine in Non-Western Cultures (2nd edition)* edited by Helaine Selin, p. 1215, Springer, ISBN 978-1-4020-4559-2.
- Sarma, Nataraja (2000). "Diffusion of Astronomy in the Ancient World". *Endeavour*. **24**: 157–164. doi: 10.1016/s0160-9327(00)01327-2[325].
- Sharma, V.N. (1995), *Sawai Jai Singh and His Astronomy*, Motilal Banarsidass, ISBN 81-208-1256-5.

- Sharma, V.N. (2008), "Observatories in India", *Encyclopaedia of the History of Science, Technology, and Medicine in Non-Western Cultures (2nd edition)* edited by Helaine Selin, pp. 1785–1788, Springer, ISBN 978-1-4020-4559-2.
- Savage-Smith, Emilie (1985), *Islamicate Celestial Globes: Their History, Construction, and Use*, Smithsonian Institution Press.
- Subbaarayappa, B.V. (1989), "Indian astronomy: an historical perspective", *Cosmic Perspectives* edited by Biswas etc., pp. 25–41. Cambridge University Press. ISBN 0-521-34354-2.
- Tripathi, V.N. (2008), "Astrology in India", *Encyclopaedia of the History of Science, Technology, and Medicine in Non-Western Cultures (2nd edition)* edited by Helaine Selin, pp. 264–267, Springer, ISBN 978-1-4020-4559-2.

Ancient Greek astronomy

Greek astronomy is astronomy written in the Greek language in classical antiquity. Greek astronomy is understood to include the ancient Greek, Hellenistic, Greco-Roman, and Late Antiquity eras. It is not limited geographically to Greece or to ethnic Greeks, as the Greek language had become the language of scholarship throughout the Hellenistic world following the conquests of Alexander. This phase of Greek astronomy is also known as **Hellenistic astronomy**, while the pre-Hellenistic phase is known as **Classical Greek astronomy**. During the Hellenistic and Roman periods, much of the Greek and non-Greek astronomers working in the Greek tradition studied at the Musaeum and the Library of Alexandria in Ptolemaic Egypt.

The development of astronomy by the Greek and Hellenistic astronomers is considered by historians to be a major phase in the history of astronomy. Greek astronomy is characterized from the start by seeking a rational, physical explanation for celestial phenomena. Most of the constellations of the northern hemisphere derive from Greek astronomy,[326] as are the names of many stars, asteroids, and planets. It was influenced by Egyptian and especially Babylonian astronomy; in turn, it influenced Indian, Arabic-Islamic and Western European astronomy.

Figure 29: *The Antikythera Mechanism was an analog computer from 150–100 BC designed to calculate the positions of astronomical objects.*

Archaic Greek astronomy

References to identifiable stars and constellations appear in the writings of Homer and Hesiod, the earliest surviving examples of Greek literature. In the *Iliad* and the *Odyssey*, Homer refers to the following celestial objects:

- the constellation Boötes
- the star cluster Hyades
- the constellation Orion
- the star cluster Pleiades
- Sirius, the Dog Star
- the constellation Ursa Major

Hesiod, who wrote in the early 7th century BC, adds the star Arcturus to this list in his poetic calendar *Works and Days*. Though neither Homer nor Hesiod set out to write a scientific work, they hint at a rudimentary cosmology of a flat earth surrounded by an "Ocean River." Some stars rise and set (disappear into the ocean, from the viewpoint of the Greeks); others are ever-visible. At certain times of the year, certain stars will rise or set at sunrise or sunset.

Speculation about the cosmos was common in Pre-Socratic philosophy in the 6th and 5th centuries BC. Anaximander (c. 610 BC–c. 546 BC) described

Figure 30: *Anaximander*

a cylindrical earth suspended in the center of the cosmos, surrounded by rings of fire. Philolaus (c. 480 BC–c. 405 BC) the Pythagorean described a cosmos with the stars, planets, Sun, Moon, Earth, and a counter-Earth (*Antichthon*)—ten bodies in all—circling an unseen central fire. Such reports show that Greeks of the 6th and 5th centuries BC were aware of the planets and speculated about the structure of the cosmos. Also, a more detailed description about the cosmos, Stars, Sun, Moon and the Earth can be found in the Orphism, which dates back to the end of the 5th century BC, and it is probably even older. Within the lyrics of the Orphic poems we can find remarkable information such as that the Earth is round, it has an axis and it moves around it in one day, it has three climate zones and that the Sun magnetizes the Stars and planets.[327]

The planets in early Greek astronomy

The name "planet" comes from the Greek term πλανήτης (*planētēs*), meaning "wanderer", as ancient astronomers noted how certain lights moved across the sky in relation to the other stars. Five planets can be seen with the naked eye: Mercury, Venus, Mars, Jupiter, and Saturn, the Greek names being Hermes, Aphrodite, Ares, Zeus and Cronus. Sometimes the luminaries, the Sun and Moon, are added to the list of naked eye planets to make a total of seven. Since the planets disappear from time to time when they approach the Sun, careful attention is required to identify all five. Observations of Venus are not straightforward. Early Greeks thought that the evening and morning appearances of Venus represented two different objects, calling it *Hesperus* ("evening star") when it appeared in the western evening sky and *Phosphorus* ("light-bringer") when it appeared in the eastern morning sky. They eventually came to recognize that both objects were the same planet. Pythagoras is given credit for this realization.

Eudoxan astronomy

In classical Greece, astronomy was a branch of mathematics; astronomers sought to create geometrical models that could imitate the appearances of celestial motions. This tradition began with the Pythagoreans, who placed astronomy among the four mathematical arts (along with arithmetic, geometry, and music). The study of number comprising the four arts was later called the *Quadrivium*.

Although he was not a creative mathematician, Plato (427–347 BC) included the *quadrivium* as the basis for philosophical education in the *Republic*. He encouraged a younger mathematician, Eudoxus of Cnidus (c. 410 BC–c. 347 BC), to develop a system of Greek astronomy. According to a modern historian of science, David Lindberg:

> *"In their work we find (1) a shift from stellar to planetary concerns, (2) the creation of a geometrical model, the "two-sphere model," for the representation of stellar and planetary phenomena, and (3) the establishment of criteria governing theories designed to account for planetary observations".*[328]

The two-sphere model is a geocentric model that divides the cosmos into two regions, a spherical Earth, central and motionless (the sublunary sphere) and a spherical heavenly realm centered on the Earth, which may contain multiple rotating spheres made of aether.

Plato's main books on cosmology are the *Timaeus* and the *Republic*. In them he described the two-sphere model and said there were eight circles or spheres carrying the seven planets and the fixed stars. According to the "Myth of Er" in the *Republic*, the cosmos is the Spindle of Necessity, attended by Sirens and spun by the three daughters of the Goddess Necessity known collectively as the Moirai or Fates.

According to a story reported by Simplicius of Cilicia (6th century), Plato posed a question for the Greek mathematicians of his day: "By the assumption of what uniform and orderly motions can the apparent motions of the planets be accounted for?" (quoted in Lloyd 1970, p. 84). Plato proposed that

Figure 31: *Renaissance woodcut illustrating the two-sphere model.*

the seemingly chaotic wandering motions of the planets could be explained by combinations of uniform circular motions centered on a spherical Earth, apparently a novel idea in the 4th century.

Eudoxus rose to the challenge by assigning to each planet a set of concentric spheres. By tilting the axes of the spheres, and by assigning each a different period of revolution, he was able to approximate the celestial "appearances." Thus, he was the first to attempt a mathematical description of the motions of the planets. A general idea of the content of *On Speeds*, his book on the planets, can be gleaned from Aristotle's *Metaphysics* XII, 8, and a commentary by Simplicius on *De caelo*, another work by Aristotle. Since all his own works are lost, our knowledge of Eudoxus is obtained from secondary sources. Aratus's poem on astronomy is based on a work of Eudoxus, and possibly also Theodosius of Bithynia's *Sphaerics*. They give us an indication of his work in spherical astronomy as well as planetary motions.

Callippus, a Greek astronomer of the 4th century, added seven spheres to Eudoxus' original 27 (in addition to the planetary spheres, Eudoxus included a sphere for the fixed stars). Aristotle described both systems, but insisted on adding "unrolling" spheres between each set of spheres to cancel the motions of the outer set. Aristotle was concerned about the physical nature of the system; without unrollers, the outer motions would be transferred to the inner planets.

Hellenistic astronomy

Planetary models and observational astronomy

The Eudoxan system had several critical flaws. One was its inability to predict motions exactly. Callippus' work may have been an attempt to correct this flaw. A related problem is the inability of his models to explain why planets appear to change speed. A third flaw is its inability to explain changes in the brightness of planets as seen from Earth. Because the spheres are concentric, planets will always remain at the same distance from Earth. This problem was pointed out in Antiquity by Autolycus of Pitane (c. 310 BC).

Apollonius of Perga (c. 262 BC–c. 190 BC) responded by introducing two new mechanisms that allowed a planet to vary its distance and speed: the eccentric deferent and the deferent and epicycle. The **deferent** is a circle carrying the planet around the Earth. (The word *deferent* comes from the Latin *ferro, ferre*, meaning "to carry.") An **eccentric deferent** is slightly off-center from Earth. In a **deferent and epicycle** model, the deferent carries a small circle, the **epicycle**, which carries the planet. The deferent-and-epicycle model can mimic the eccentric model, as shown by **Apollonius' theorem**. It can also explain retrogradation, which happens when planets appear to reverse their motion through the zodiac for a short time. Modern historians of astronomy have determined that Eudoxus' models could only have approximated retrogradation crudely for some planets, and not at all for others.

In the 2nd century BC, Hipparchus, aware of the extraordinary accuracy with which Babylonian astronomers could predict the planets' motions, insisted that Greek astronomers achieve similar levels of accuracy. Somehow he had access to Babylonian observations or predictions, and used them to create better geometrical models. For the Sun, he used a simple eccentric model, based on observations of the equinoxes, which explained both changes in the speed of the Sun and differences in the lengths of the seasons. For the Moon, he used a deferent and epicycle model. He could not create accurate models for the remaining planets, and criticized other Greek astronomers for creating inaccurate models.

Hipparchus also compiled a star catalogue. According to Pliny the Elder, he observed a *nova* (new star). So that later generations could tell whether other stars came to be, perished, moved, or changed in brightness, he recorded the position and brightness of the stars. Ptolemy mentioned the catalogue in connection with Hipparchus' discovery of precession. (Precession of the equinoxes is a slow motion of the place of the equinoxes through the zodiac, caused by the shifting of the Earth's axis). Hipparchus thought it was caused by the motion of the sphere of fixed stars.

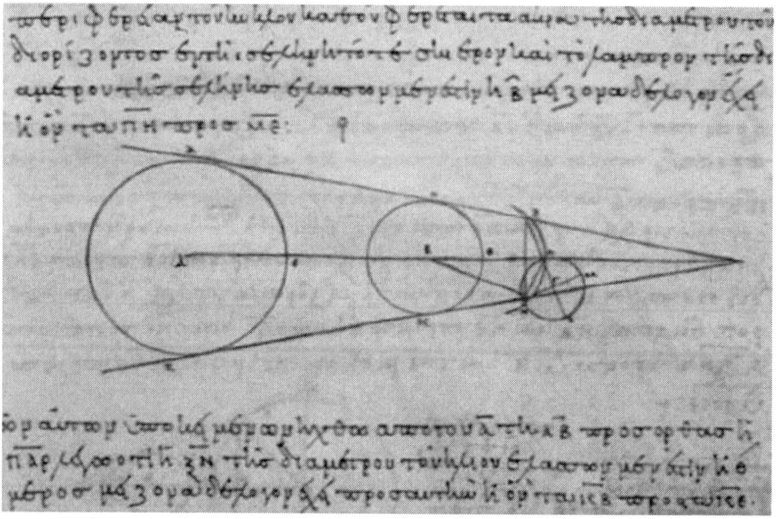

Figure 32: *Aristarchus's 3rd-century BCE calculations on the relative sizes of (from left) the Sun, Earth and Moon, from a 10th-century CE Greek copy*

Heliocentrism and cosmic scales

In the 3rd century BC, Aristarchus of Samos proposed an alternate cosmology (arrangement of the universe): a heliocentric model of the solar system, placing the Sun, not the Earth, at the center of the known universe (hence he is sometimes known as the "Greek Copernicus"). His astronomical ideas were not well-received, however, and only a few brief references to them are preserved. We know the name of one follower of Aristarchus: Seleucus of Seleucia.

Aristarchus also wrote a book *On the Sizes and Distances of the Sun and Moon*, which is his only work to have survived. In this work, he calculated the sizes of the Sun and Moon, as well as their distances from the Earth in Earth radii. Shortly afterwards, Eratosthenes calculated the size of the Earth, providing a value for the Earth radii which could be plugged into Aristarchus' calculations. Hipparchus wrote another book *On the Sizes and Distances of the Sun and Moon*, which has not survived. Both Aristarchus and Hipparchus drastically underestimated the distance of the Sun from the Earth.

Astronomy in the Greco-Roman and Late Antique eras

Hipparchus is considered to have been among the most important Greek astronomers, because he introduced the concept of exact prediction into astronomy. He was also the last innovative astronomer before Claudius Ptolemy, a mathematician who worked at Alexandria in Roman Egypt in the 2nd century. Ptolemy's works on astronomy and astrology include the *Almagest*, the *Planetary Hypotheses*, and the *Tetrabiblos*, as well as the *Handy Tables*, the *Canobic Inscription*, and other minor works.

Ptolemaic astronomy

The *Almagest* is one of the most influential books in the history of Western astronomy. In this book, Ptolemy explained how to predict the behavior of the planets, as Hipparchus could not, with the introduction of a new mathematical tool, the equant. The *Almagest* gave a comprehensive treatment of astronomy, incorporating theorems, models, and observations from many previous mathematicians. This fact may explain its survival, in contrast to more specialized works that were neglected and lost. Ptolemy placed the planets in the order that would remain standard until it was displaced by the heliocentric system and the Tychonic system:

1. Moon
2. Mercury
3. Venus
4. Sun
5. Mars
6. Jupiter
7. Saturn
8. Fixed stars

The extent of Ptolemy's reliance on the work of other mathematicians, in particular his use of Hipparchus' star catalogue, has been debated since the 19th century. A controversial claim was made by Robert R. Newton in the 1970s. in *The Crime of Claudius Ptolemy*, he argued that Ptolemy faked his observations and falsely claimed the catalogue of Hipparchus as his own work. Newton's theories have not been adopted by most historians of astronomy.

A few mathematicians of Late Antiquity wrote commentaries on the *Almagest*, including Pappus of Alexandria as well as Theon of Alexandria and his daughter Hypatia. Ptolemaic astronomy became standard in medieval western European and Islamic astronomy until it was displaced by Maraghan, heliocentric and Tychonic systems by the 16th century. However, recently discovered

Figure 33: *Greek equatorial sun dial, Ai-Khanoum, Afghanistan 3rd-2nd century BC.*

manuscripts reveal that Greek astrologers of Antiquity continued using pre-Ptolemaic methods for their calculations (Aaboe, 2001).

Influence on Indian astronomy

Hellenistic astronomy is known to have been practiced near India in the Greco-Bactrian city of Ai-Khanoum from the 3rd century BC. Various sun-dials, including an equatorial sundial adjusted to the latitude of Ujjain have been found in archaeological excavations there.[329] Numerous interactions with the Mauryan Empire, and the later expansion of the Indo-Greeks into India suggest that some transmission may have happened during that period.[330]

Several Greco-Roman astrological treatises are also known to have been imported into India during the first few centuries of our era. The *Yavanajataka* ("Sayings of the Greeks") was translated from Greek to Sanskrit by Yavanesvara during the 2nd century, under the patronage of the Western Satrap Saka king Rudradaman I. Rudradaman's capital at Ujjain "became the Greenwich of Indian astronomers and the Arin of the Arabic and Latin astronomical treatises; for it was he and his successors who encouraged the introduction of Greek horoscopy and astronomy into India."

Later in the 6th century, the *Romaka Siddhanta* ("Doctrine of the Romans"), and the *Paulisa Siddhanta* ("Doctrine of Paul") were considered as two of the five main astrological treatises, which were compiled by Varahamihira in his *Pañca-siddhāntikā* ("Five Treatises").[331] Varahamihira wrote in the Brihat-Samhita: "For, the Greeks are foreigners. This science is well established among them. Although they are revered as sages, how much more so is a twice-born person who knows the astral science."[332]

Sources for Greek astronomy

Many Greek astronomical texts are known only by name, and perhaps by a description or quotations. Some elementary works have survived because they were largely non-mathematical and suitable for use in schools. Books in this class include the *Phaenomena* of Euclid and two works by Autolycus of Pitane. Three important textbooks, written shortly before Ptolemy's time, were written by Cleomedes, Geminus, and Theon of Smyrna. Books by Roman authors like Pliny the Elder and Vitruvius contain some information on Greek astronomy. The most important primary source is the *Almagest*, since Ptolemy refers to the work of many of his predecessors (Evans 1998, p. 24).

Famous astronomers of antiquity

In addition to the authors named in the article, the following list of people who worked on mathematical astronomy or cosmology may be of interest.

- Anaxagoras
- Archimedes
- Archytas
- Aristaeus
- Aristarchus
- Aristyllus
- Conon of Samos
- Democritus
- Empedocles
- Heraclides Ponticus
- Hicetas
- Hippocrates of Chios
- Macrobius
- Martianus Capella
- Menelaus of Alexandria (Menelaus theorem)
- Meton of Athens
- Parmenides

- Porphyry
- Posidonius
- Proclus
- Thales
- Theodosius of Bithynia

References

- Aaboe, Asger H. (2001). *Episodes from the Early History of Astronomy*. New York: Springer. ISBN 978-0-387-95136-2.
- Dreyer, John L. E. (1953). *A History of Astronomy from Thales to Kepler* (2nd ed.). New York: Dover Publications. ISBN 978-0-486-60079-6.
- Evans, James (1998). *The History and Practice of Ancient Astronomy*. New York: Oxford University Press. ISBN 978-0-19-509539-5.
- Heath, Thomas L. (1913). *Aristarchus of Samos*. Oxford: Clarendon Press.
- Lindberg, David C. (2010). *The Beginnings of Western Science: The European Scientific Tradition in Philosophical, Religious, and Institutional Context, 600 B.C. to A.D. 1450* (2 ed.). Chicago: University of Chicago Press. ISBN 978-0-226-48204-0.
- Lloyd, Geoffrey E. R. (1970). *Early Greek Science: Thales to Aristotle*. New York: W. W. Norton & Co.
- Neugebauer, Otto E. (1975). *A History of Ancient Mathematical Astronomy*. Berlin: Springer. ISBN 978-0-387-06995-1.
- Newton, Robert R. (1977). *The Crime of Claudius Ptolemy*. Baltimore: Johns Hopkins University Press. ISBN 978-0-8018-1990-2.
- Pedersen, Olaf (1993). *Early Physics and Astronomy: A Historical Introduction* (2nd ed.). Cambridge: Cambridge University Press. ISBN 978-0-521-40340-5.
- Revello, Manuela (2013). *"Sole, luna ed eclissi in Omero", in TECHNAI 4*. Pisa-Roma: Fabrizio Serra editore. pp. 13–32.
- Toomer, Gerald J. (1998). *Ptolemy's Almagest*. Princeton: Princeton University Press. ISBN 978-0-691-00260-6.

External links

- Almagest Planetary Model Animations[333]
- MacTutor History of Mathematics Archive[334]

Egyptian astronomy

Egyptian astronomy begins in prehistoric times, in the Predynastic Period. In the 5th millennium BCE, the stone circles at Nabta Playa may have made use of astronomical alignments. By the time the historical Dynastic Period began in the 3rd millennium BCE, the 365-day period of the Egyptian calendar was already in use, and the observation of stars was important in determining the annual flooding of the Nile. The Egyptian pyramids were carefully aligned towards the pole star, and the temple of Amun-Re at Karnak was aligned on the rising of the midwinter sun. Astronomy played a considerable part in fixing the dates of religious festivals and determining the hours of the night, and temple astrologers were especially adept at watching the stars and observing the conjunctions, phases, and risings of the sun, moon and planets.

In Ptolemaic Egypt, the Egyptian tradition merged with Greek astronomy and Babylonian astronomy, with the city of Alexandria in Lower Egypt becoming the centre of scientific activity across the Hellenistic world. Roman Egypt produced the greatest astronomer of the era, Ptolemy (90-168 CE). His works on astronomy, including the Almagest, became the most influential books in the history of Western astronomy. Following the Muslim conquest of Egypt, the region came to be dominated by Arabic culture and Islamic astronomy. The astronomer Ibn Yunus (c. 950-1009) observed the sun's position for many years using a large astrolabe, and his observations on eclipses were still used centuries later. In 1006, Ali ibn Ridwan observed the SN 1006, a supernova regarded as the brightest stellar event in recorded history, and left the most detailed description of it. In the 14th century, Najm al-Din al-Misri wrote a treatise describing over 100 different types of scientific and astronomical instruments, many of which he invented himself. In the 20th century, Farouk El-Baz from Egypt worked for NASA and was involved in the first Moon landings with the Apollo program, where he assisted in the planning of scientific explorations of the Moon.

Ancient Egypt

Egyptian astronomy begins in prehistoric times. The presence of stone circles at Nabta Playa in Upper Egypt dating from the 5th millennium BCE show the importance of astronomy to the religious life of ancient Egypt even in the prehistoric period. The annual flooding of the Nile meant that the heliacal risings, or first visible appearances of stars at dawn, were of special interest in determining when this might occur, and it is no surprise that the 365-day period of the Egyptian calendar was already in use at the beginning of Egyptian history. The constellation system used among the Egyptians also appears to have been essentially of native origin.

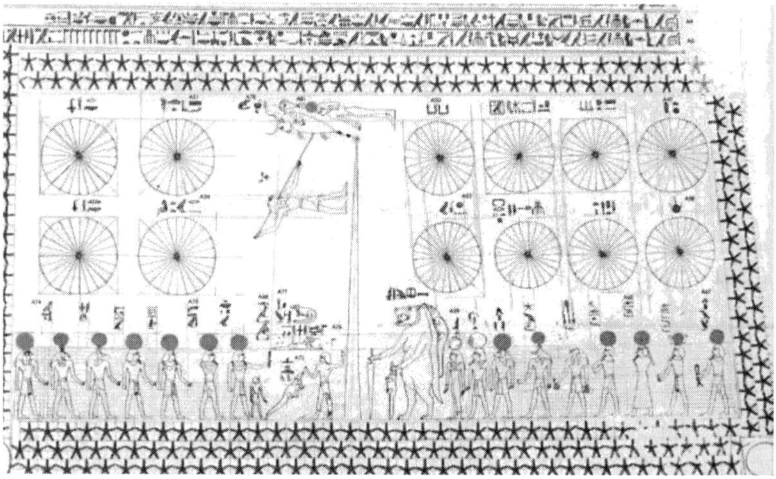

Figure 34: *Chart from Senemut's tomb, 18th dynasty*[335]

Figure 35: *Nut, Egyptian goddess of the sky, with the star chart in the tomb of Ramses VI*

Figure 36: *Plan of a stone circle at Nabta, Egypt*

The precise orientation of the Egyptian pyramids serves as a lasting demonstration of the high degree of technical skill in watching the heavens attained in the 3rd millennium BCE. It has been shown the pyramids were aligned towards the pole star, which, because of the precession of the equinoxes, was at that time Thuban, a faint star in the constellation of Draco.[336] Evaluation of the site of the temple of Amun-Re at Karnak, taking into account the change over time of the obliquity of the ecliptic, has shown that the Great Temple was aligned on the rising of the midwinter sun.[337] The length of the corridor down which sunlight would travel would have limited illumination at other times of the year.

Astronomy played a considerable part in religious matters for fixing the dates of festivals and determining the hours of the night. The titles of several temple books are preserved recording the movements and phases of the sun, moon and stars. The rising of Sirius (Egyptian: Sopdet, Greek: Sothis) at the beginning of the inundation was a particularly important point to fix in the yearly calendar. One of the most important Egyptian astronomical texts was the Book of Nut, going back to the Middle Kingdom or earlier.

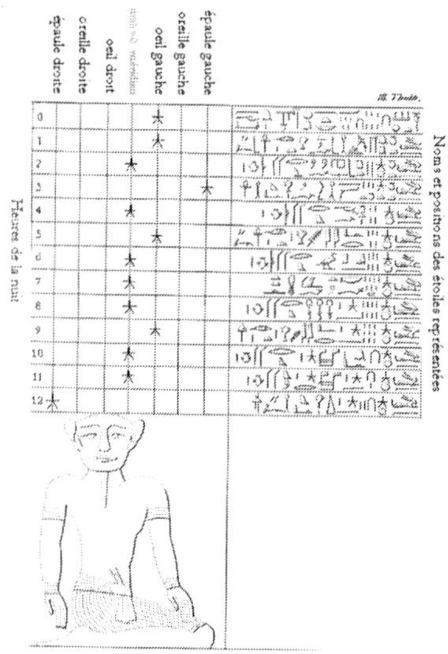

Figure 37: *'Star clock' method from the tomb of Rameses VI*

The First Intermediate Period

Beginning with the 9th Dynasty, ancient Egyptians produced 'Diagonal star tables', which were usually painted on the inside surface of wooden coffin lids.[338] This practice continued until the 12th dynasty.[339] These 'Diagonal star tables' or star charts are also known as 'diagonal star clocks'; in the past they have also been known as 'star calendars', or 'decanal clocks'.[340] These star charts featuring the paintings of Egyptian deities, decans, constellations, and star observations are also found on the ceilings of tombs and temples.

From the tables of stars on the ceiling of the tombs of Rameses VI and Rameses IX it seems that for fixing the hours of the night a man seated on the ground faced the Astrologer in such a position that the line of observation of the pole star passed over the middle of his head. On the different days of the year each hour was determined by a fixed star culminating or nearly culminating in it, and the position of these stars at the time is given in the tables as in the centre, on the left eye, on the right shoulder, etc. According to the texts, in founding or rebuilding temples the north axis was determined by the same apparatus, and we may conclude that it was the usual one for astronomical observations. In careful hands, it might give results of a high degree of accuracy.

Macrobius Ambrosius Theodosius (*floruit* 395–423 CE) attributed the planetary theory where the Earth rotates on its axis and the interior planets Mercury

Figure 38: *An Egyptian 30th-dynasty (Ptolemaic) terracotta astrological disc at the Los Angeles County Museum of Art.*

and Venus revolve around the Sun which in turn revolves around the Earth, to the ancient Egyptians. He named it the "Egyptian System," and stated that "it did not escape the skill of the Egyptians," though there is no other evidence it was known in ancient Egypt.

Greco-Roman Egypt

Writing in the Roman era, Clement of Alexandria gives some idea of the importance of astronomical observations to the sacred rites:

> *And after the Singer advances the Astrologer (ὡροσκόπος), with a horologium (ὡρολόγιον) in his hand, and a palm (φοίνιξ), the symbols of astrology. He must know by heart the Hermetic astrological books, which are four in number. Of these, one is about the arrangement of the fixed stars that are visible; one on the positions of the sun and moon and five planets; one on the conjunctions and phases of the sun and moon; and one concerns their risings.*[341]

The astrologer's instruments (*horologium* and *palm*) are a plumb line and sighting instrument. They have been identified with two inscribed objects in the Berlin Museum; a short handle from which a plumb line was hung, and a palm branch with a sight-slit in the broader end. The latter was held close to the eye,

Figure 39: *Astronomical ceiling relief from Dendera, Egypt*

the former in the other hand, perhaps at arms length. The "Hermetic" books which Clement refers to are the Egyptian theological texts, which probably have nothing to do with Hellenistic Hermetism.[342]

Following Alexander the Great's conquests and the foundation of Ptolemaic Egypt, the native Egyptian tradition of astronomy had merged with Greek astronomy as well as Babylonian astronomy. The city of Alexandria in Lower Egypt became the centre of scientific activity throughout the Hellenistic civilization. The greatest Alexandrian astronomer of this era was the Greek, Eratosthenes (c. 276-195 BCE), who calculated the size of the Earth, providing an estimate for the circumference of the Earth.

Following the Roman conquest of Egypt, the region once again became the centre of scientific activity throughout the Roman Empire. The greatest astronomer of this era was the Hellenized Egyptian, Ptolemy (90-168 CE). Originating from the Thebaid region of Upper Egypt, he worked at Alexandria and wrote works on astronomy including the *Almagest*, the *Planetary Hypotheses*, and the *Tetrabiblos*, as well as the *Handy Tables*, the *Canobic Inscription*, and other minor works. The *Almagest* is one of the most influential books in the history of Western astronomy. In this book, Ptolemy explained how to predict the behavior of the planets with the introduction of a new mathematical tool, the equant.

A few mathematicians of late Antiquity wrote commentaries on the *Almagest*, including Pappus of Alexandria as well as Theon of Alexandria and his daughter Hypatia. Ptolemaic astronomy became standard in medieval western European and Islamic astronomy until it was displaced by Maraghan, heliocentric and Tychonic systems by the 16th century.

Arabic-Islamic Egypt

Following the Muslim conquest of Egypt, the region came to be dominated by Arabic culture. It was ruled by the Rashidun, Umayyad and Abbasid Caliphates up until the 10th century, when the Fatimids founded their own Caliphate centred around the city of Cairo in Egypt. The region once again became a centre of scientific activity, competing with Baghdad for intellectual dominance in the medieval Islamic world. By the 13th century, the city of Cairo eventually overtook Baghdad as the intellectual center of the Islamic world.Wikipedia:Citation needed

Ibn Yunus (c. 950-1009) observed more than 10,000 entries for the sun's position for many years using a large astrolabe with a diameter of nearly 1.4 meters. His observations on eclipses were still used centuries later in Simon Newcomb's investigations on the motion of the moon, while his other observations inspired Laplace's *Obliquity of the Ecliptic* and *Inequalities of Jupiter and Saturn*.Wikipedia:Please clarify In 1006, Ali ibn Ridwan observed the supernova of 1006, regarded as the brightest stellar event in recorded history, and left the most detailed description of the temporary star. He says that the object was two to three times as large as the disc of Venus and about one-quarter the brightness of the Moon, and that the star was low on the southern horizon.

The astrolabic quadrant was invented in Egypt in the 11th century or 12th century, and later known in Europe as the "Quadrans Vetus" (Old Quadrant). In 14th century Egypt, Najm al-Din al-Misri (c. 1325) wrote a treatise describing over 100 different types of scientific and astronomical instruments, many of which he invented himself.

In the 20th century, Farouk El-Baz from Egypt worked for NASA and was involved in the first Moon landings with the Apollo program, where he was secretary of the *Landing Site Selection Committee*, *Principal Investigator of Visual Observations and Photography*, chairman of the *Astronaut Training Group*, and assisted in the planning of scientific explorations of the Moon, including the selection of landing sites for the Apollo missions and the training of astronauts in lunar observations and photography.

References

- King, David A. (2004), "Reflections on some new studies on applied science in Islamic societies (8th-19th centuries)", *Islam & Science*, June 2004.
- King, David A.; Cleempoel, Koenraad Van; Moreno, Roberto (2002), "A Recently Discovered Sixteenth-Century Spanish Astrolabe", *Annals of Science*, **59** (4): 331–362, doi: 10.1080/00033790110095813[343]

Further reading

- Marshall Clagett, (2004), *Ancient Egyptian Science: A Source Book. Volume Two: Calendars, Clocks, and Astronomy*, American Philosophical Society, ISBN 0-87169-214-7.
- Massimiliano Franci, *Astronomia egizia, Introduzione alle conoscenze astronomiche dell'antico Egitto*, Edarc, Firenze 2010, ISBN 978-88-86428-94-1.

External links

- Media related to Ancient Egyptian astronomy at Wikimedia Commons.
- Symons, S.L., Cockcroft, R., Bettencourt, J. and Koykka, C., 2013. Ancient Egyptian Astronomy. [Online database] Available at: <http://aea.physics.mcmaster.ca/>.

Chinese astronomy

Astronomy in China has a very long history, with historians indicating that the Chinese were the most persistent and accurate observers of celestial phenomena anywhere in the world before the Arabs.[344] Star names later categorized in the twenty-eight mansions have been found on oracle bones unearthed at Anyang, dating back to the middle Shang Dynasty (Chinese Bronze Age), and the mansion (xiù:宿) system's nucleus seems to have taken shape by the time of the ruler Wu Ding (1339-1281 BC).[345]

Detailed records of astronomical observations began during the Warring States period (fourth century BC) and flourished from the Han period onward. Chinese astronomy was equatorial, centered as it was on close observation of circumpolar stars, and was based on different principles from those prevailing in traditional Western astronomy, where heliacal risings and settings of zodiac constellations formed the basic ecliptic framework.[346]

Some elements of Indian astronomy reached China with the expansion of Buddhism after the Eastern Han Dynasty (25–220 AD), but the most detailed incorporation of Indian astronomical thought occurred during the Tang Dynasty (618-907), when numerous Indian astronomers took up residence in the Chinese capital, and Chinese scholars, such as the great Tantric Buddhist monk and mathematician Yi Xing, mastered its system. Islamic astronomers collaborated closely with their Chinese colleagues during the Yuan Dynasty, and, after a period of relative decline during the Ming Dynasty, astronomy was revitalized under the stimulus of Western cosmology and technology after the

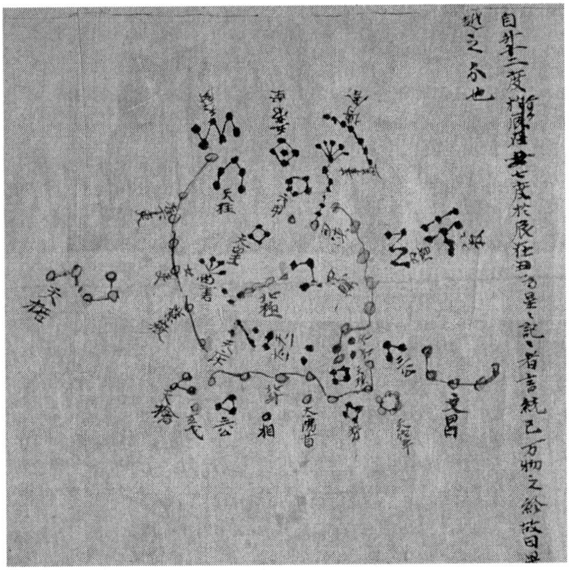

Figure 40: *The Dunhuang map from the Tang Dynasty (North Polar region). This map is thought to date from the reign of Emperor Zhongzong of Tang (705–710). Founded in Dunhuang, Gansu. Constellations of the three schools were distinguished with different colors: white, black and yellow for stars of Wu Xian, Gan De and Shi Shen respectively. The whole set of star maps contained 1,300 stars.*

Jesuits established their missions. The telescope was introduced in the seventeenth century. In 1669, the Peking observatory was completely redesigned and refitted under the direction of Ferdinand Verbiest. Today, China continues to be active in astronomy, with many observatories and its own space program.

Early history

History of science and technology in China

- Inventions
  - Four Great Inventions
- Discoveries

By subject

• Mathematics
• Astronomy
• Calendar
• Units of measurement
• Cartography
• Printing
• Ceramics
• Metallurgy
• Coinage
• Chinese Alchemy
• Traditional medicine
  • herbology
• Agriculture
  • Sericulture
• Silk industry
• Architecture
  • Classic gardens
  • Bridges
• Transport
  • Navigation
• Military
  • Navy

By era
• Han • Tang • Song • People's Republic   • Agriculture
• v • t • e[347]

Purpose of astronomical observations in the past

One of the main functions was for the purpose of timekeeping. The Chinese used a lunisolar calendar, but, because the cycles of the sun and the moon are different, intercalation had to be done.

The Chinese calendar was considered to be a symbol of a dynasty. As dynasties would rise and fall, astronomers and astrologers of each period would often prepare a new calendar to be made, with observations for that purpose.

Astrological divination was also an important part of astronomy. Astronomers took careful note of guest stars, which suddenly appeared among the fixed stars. The supernova that created the Crab Nebula observed in 1054, now known as the SN 1054, is an example of a guest star observed by Chinese astronomers, recorded also by the Arab astronomers, although it was not

Figure 41: *Wide view of the Crab Nebula.*

recorded by their European contemporaries. Ancient astronomical records of phenomena like comets and supernovae are sometimes used in modern astronomical studies.

Indian influence

Indian astronomy reached China with the expansion of Buddhism during the Later Han (25–220 CE). Further translation of Indian works on astronomy was completed in China by the Three Kingdoms era (220–265 CE). However, the most detailed incorporation of Indian astronomy occurred only during the Tang Dynasty (618–907 CE) when a number of Chinese scholars—such as Yi Xing— were versed both in Indian and Chinese astronomy. A system of Indian astronomy was recorded in China as *Jiuzhi-li* (718 CE), the author of which was an Indian by the name of Qutan Xida—a translation of Devanagari Gotama Siddha—the director of the Tang dynasty's national astronomical observatory.During the 8th century, the astronomical table of sines by the Indian astronomer and mathematician, Aryabhatta (476-550), were translated into the Chinese astronomical and mathematical book of the *Treatise on Astrology of the Kaiyuan Era* (*Kaiyuan Zhanjing*), compiled in 718 CE during the Tang Dynasty.[348] The *Kaiyuan Zhanjing* was compiled by Gautama Siddha, an astronomer and astrologer born in Chang'an, and whose family was originally from India. He was also notable for his translation of the Nava-graha calendar into Chinese.[349] Gautama Siddha introduced Indian numerals with zero (○) in 718 in China as a replacement of counting rods. In 3rd-century C.E, the Matanaga avadha was translated into Chinese.although the original is believed to date earlier. It gives the lengths of monthly shadows of a 12-inch gnomon, which is the standard parameter of Indian astronomy.The work also mentions the 28 Indian nakshatras.[350] In the beginning of the second century, Sardulakarnavadana was translated into Chinese several times, This work contains the usual Sanskrit names of the 28 nakshatras. starting with krttika.[351] From the 1st century onward Lalitavistara was translated into Chinese several times. It is in this work that the famous Buddhist centesimal-scale counting occurs during the dialogue between Prince Gautamaand and the mathematician Arjuna. The first series of counts ends with tallaksana (= 10^{53}), beyond which eight more ganana series are mentioned.Atomic-scale counting is also mentioned[352].The Mahaprajnaparamita Sastra (of Nagarjuna, second century) was translated into Chinese by Kumarajiva in the early fifth century. 16 The astronomical parameters mentioned in this translation are comparable to those given in the Vedanga Jyotisha.Indian system of numeration appeared in the Chinese work Ta PaoChi Ching (Maharatnakuta Sutra), translated by Upasunya (in 541 c.e.) The Chinese translations of the following works are mentioned in the Sui Shu, or Official History of the Sui Dynasty (seventh century):

- Po-lo-men Thien Wen Ching (Brahminical Astronomical Classic) in 21 books.
- Po-lo-men Chieh-Chhieh Hsien-jen Thien Wen Shuo (Astronomical Theories of

Brahman.a Chieh-Chhieh Hsienjen) in 30 books.

- Po-lo-men Thien Ching (Brahminical Heavenly Theory) in one book.
- Mo-teng-Chia Ching Huang-thu (Map of Heaven in the Matangi Sutra) in one

book.

- Po-lo-men Suan Ching (Brahminical Arithmetical Classic) in three books.
- Po-lo-men Suan Fa (Brahminical Arithmetical Rules) in one book.
- Po-lo-men Ying Yang Suan Ching (Brahminical Method of Calculating Time)

Although these translations are lost, they were also mentioned in other sources.

Cosmology

The Chinese developed three different cosmological models. The Gai Tian, or hemispherical dome, model conceived the heavens as a hemisphere lying over a dome-shaped earth. The second cosmological model, associated with the Hun Tian school, saw the heavens as a celestial sphere not unlike the spherical models developed in the Greek and Hellenistic traditions. The third cosmology, associated with the Xuan Ye school, viewed the heavens as infinite in extent and the celestial bodies as floating about at rare intervals, and "the speed of the luminaries depends on their individual natures, which shows they are not attached to anything."

Constellations

The divisions of the sky began with the Northern Dipper and the 28 mansions.

In 1977, a lacquer box was excavated from the tomb of Yi, the marquis of Zeng, in Suixian, Hubei Province. Names of the 28 lunar mansions were found on the cover of the box, proving that the use of this classification system was made before 433 BC.

As lunar mansions have such an ancient origin, the meanings of most of their names have become obscure. Even worse, the name of each lunar mansion consists of only one Chinese word, the meaning of which could vary at different times in history. The meanings of the names are still under discussion.

Besides the 28 lunar mansions, most constellations are based on the works of Shi Shen-fu and Gan De, who were astrologists during the period of Warring States (481 BC - 221 BC) in China.

In the late period of the Ming Dynasty, the agricultural scientist and mathematician Xu Guangqi (1562 - 1633 AD) introduced 23 additional constellations near to the Celestial South Pole, which are based on star catalogues from the West (see Matteo Ricci).

Star catalogues and maps

Star catalogues

In the fourth century BC, the two Chinese astronomers responsible for the earliest information going into the star catalogues were Shi Shen and Gan De of the Warring States period.[353]

Author	Translated name	Chinese catalogue name	Pinyin
Shi Shen	Shi Shen astronomy	石申天文	Shi Shen tianwen
Gan De	Astronomic star observation	天文星占	Tianwen xingzhan

These books appeared to have lasted until the sixth century, but were lost after that. A number of books share similar names, often quoted and named after them. These texts should *not* be confused with the original catalogues written by them. Notable works that helped preserve the contents include:

Author	Translated name	Chinese name	Pinyin	Comments
Sima Qian	Book of Celestial Offices	天官書	Tianguan shu	This is the astronomical chapter of the *Records of the Grand Historian*, a massive history compiled during the late 2nd century BC by the Han-era scholar and official Sima Qian. This chapter provides a star catalogue and discusses the schools of Gan De and Shi Shen.[354]
Ma Xian (馬顯)	Star Manual of the Masters Gan and Shi	甘石星經	Gan Shi Xingjing	Despite having the name credited to Shi and Gan, it was lost and later compiled circa 579 AD as an appendix to the Treatise on Astrology of the Kaiyuan Era, and summarized in the book 郡齋讀書志.[355]
	Book of Jin	晉書	Jin shu	In the astronomical chapters of the text
	Book of Sui	隋書	Sui shu	
Gautama Siddha	Treatise on Astrology of the Kaiyuan Era	開元占經	Kaiyuan Zhanjing	During the reign of Emperor Xuanzong of Tang (712-756 AD). After analyzing and providing a summary on the work of Gan De and Shi Shen, Tang era astronomers mentioned the names of more than 800 stars that were found, 121 of them marked with positions.[356] The astronomical table of sines by the Indian astronomer and mathematician Aryabhata were also translated into the *Kaiyuan Zhanjing*.
	The Great Firmament Star Manual Common to Astrology	通占大象曆星經	Tongzhan taxiangli xingjing	This renamed star manual is incorporated in the Taoist book *Daozang*.

Wu Xian (巫咸) has been one of the astronomers in debate. He is often represented as one of the "Three Schools Astronomical tradition" along with Gan and Shi.[357] The Chinese classic text *Star Manual of Master Wu Xian* (巫咸星經) and its authorship is still in dispute, because it mentioned names of twelve countries that did not exist in the Shang Dynasty, the era of which it was supposed to have been written. Moreover, it was customary in the past for the Chinese to forge works of notable scholars, as this could lead to a possible

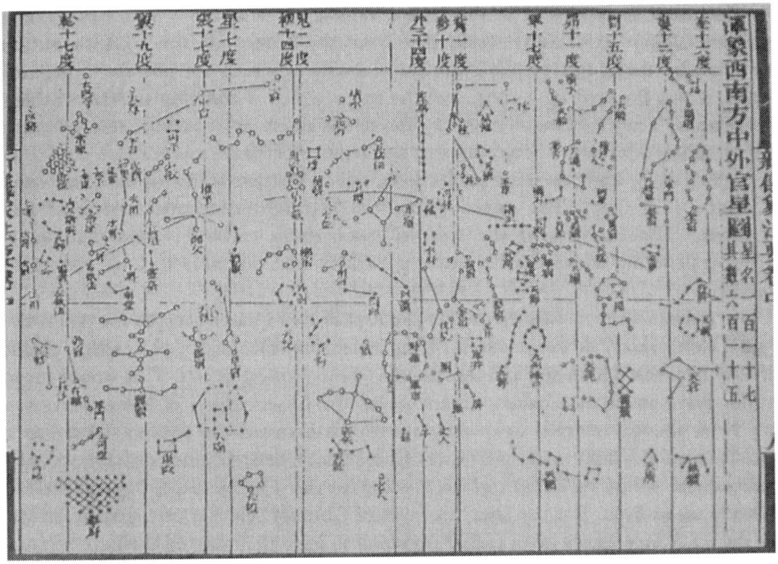

Figure 42: *A star map with a cylindrical projection. Su Song's star maps represent the oldest existent ones in printed form.*

explanation for the inconsistencies found. Wu Xian is generally mentioned as the astronomer who lived many years before Gan and Shi.

The Han Dynasty astronomer and inventor Zhang Heng (78-139 AD) not only catalogued some 2500 different stars, but also recognized more than 100 different constellations. Zhang Heng also published his work *Ling Xian*, a summary of different astronomical theories in China at the time. In the subsequent period of the Three Kingdoms (220-280 AD), Chen Zhuo (陳卓) combined the work of his predecessors, forming another star catalogue. This time, 283 constellations and 1464 stars were listed. The astronomer Guo Shoujin of the Yuan Dynasty (1279-1368 AD) created a new catalogue, which was believed to contain thousands of stars. Unfortunately, many of the documents of that period were destroyed, including that of Shoujin. *Imperial Astronomical Instruments* (儀象考成) was published in 1757 and contains 3083 stars exactly.

Star maps

The Chinese drew many maps of stars in the past centuries. It is debatable as to which counts as the oldest star maps, since pottery and old artifacts can also be considered star maps. One of the oldest existent star maps in printed form is from Su Song's (1020-1101 AD) celestial atlas of 1092 AD, which was included in the horological treatise on his clocktower. The most famous one is perhaps the Dunhuang map found in Dunhuang, Gansu. Uncovered by the British archaeologist Marc Aurel Stein in 1907, the star map was brought to the British Museum in London. The map was drawn on paper and represents the complete sky, with more than 1,350 stars. Although ancient Babylonians and Greeks also observed the sky and catalogued stars, no such complete record of the stars may exist or survive. Hence, this is the oldest chart of the skies at present.

According to recent studies, the map may date the manuscript to as early as the seventh century AD (Tang Dynasty). Scholars believe the star map dating from 705 to 710 AD, which is the reign of Emperor Zhongzong of Tang. There are some texts (Monthly Ordinances, 月令) describing the movement of the sun among the sky each month, which was not based on the observation at that time.

Solar and lunar eclipses

Chinese astronomers recorded 1,600 observations of solar and lunar eclipses from 750 BC.[358] The ancient Chinese astronomer Shi Shen (fl. fourth century BC) was aware of the relation of the moon in a solar eclipse, as he provided instructions in his writing to predict them by using the relative positions of the moon and the sun.[359] The radiating-influence theory, where the moon's light was nothing but a reflection of the sun's, was supported by the mathematician and music theorist Jing Fang (78–37 BC), yet opposed by the Chinese philosopher Wang Chong (27–97 AD), who made clear in his writing that this theory was nothing new.[360] Jing Fang wrote:

> *The moon and the planets are Yin; they have shape but no light. This they receive only when the sun illuminates them. The former masters regarded the sun as round like a crossbow bullet, and they thought the moon had the nature of a mirror. Some of them recognized the moon as a ball too. Those parts of the moon which the sun illuminates look bright, those parts which it does not, remain dark.*[361]

The ancient Greeks had known this as well, since Parmenides and Aristotle supported the theory of the moon shining because of reflected light. The Chinese astronomer and inventor Zhang Heng (78–139 AD) wrote of both solar eclipse and lunar eclipse in the publication of *Ling Xian* (靈憲), 120 AD:

> *The sun is like fire and the moon like water. The fire gives out light and the water reflects it. Thus the moon's brightness is produced from the radiance of the sun, and the moon's darkness (pho) is due to (the light of) the sun being obstructed (pi). The side which faces the sun is fully lit, and the side*

> *which is away from it is dark. The planets (as well as the moon) have the nature of water and reflect light. The light pouring forth from the sun (tang jih chih chhung kuang) does not always reach the moon owing to the obstruction (pi) of the earth itself—this is called 'an-hsü', a* **lunar eclipse**. *When (a similar effect) happens with a planet (we call it) an occulation (hsing wei); when the moon passes across (kuo) (the sun's path) then there is a* **solar eclipse** *(shih).*[362]

The later Song Dynasty scientist Shen Kuo (1031–1095) used the models of lunar eclipse and solar eclipse in order to prove that the celestial bodies were round, not flat. This was an extension of the reasoning of Jing Fang and other theorists as early as the Han Dynasty. In his *Dream Pool Essays* of 1088 AD, Shen related a conversation he had with the director of the Astronomical Observatory, who had asked Shen if the shapes of the sun and the moon were round like balls or flat like fans. Shen Kuo explained his reasoning for the former:

> *If they were like balls they would surely obstruct each other when they met. I replied that these celestial bodies were certainly like balls. How do we know this? By the waxing and waning of the moon. The moon itself gives forth no light, but is like a ball of silver; the light is the light of the sun (reflected). When the brightness is first seen, the sun (-light passes almost) alongside, so the side only is illuminated and looks like a crescent. When the sun gradually gets further away, the light shines slanting, and the moon is full, round like a bullet. If half of a sphere is covered with (white) powder and looked at from the side, the covered part will look like a crescent; if looked at from the front, it will appear round. Thus we know that the celestial bodies are spherical.*

When he asked Shen Kuo why eclipses occurred only on an occasional basis while in conjunction and opposition once a day, Shen Kuo wrote:

> *I answered that the ecliptic and the moon's path are like two rings, lying one over the other, but distant by a small amount. (If this obliquity did not exist), the sun would be eclipsed whenever the two bodies were in conjunction, and the moon would be eclipsed whenever they were exactly in position. But (in fact) though they may occupy the same degree, the two paths are not (always) near (each other), and so naturally the bodies do not (intrude) upon one another.*[363]

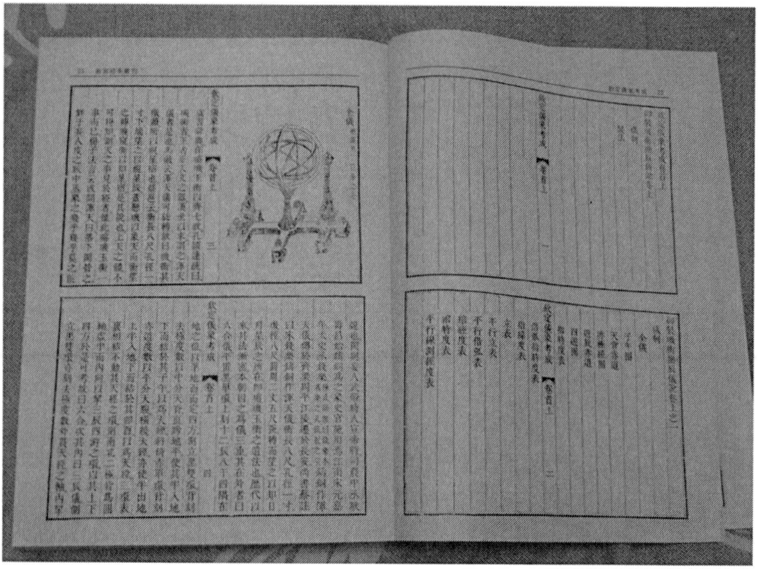

Figure 43: *A method of making observation instruments at the times of Qing Dynasty*

Equipment and innovation

Armillary sphere (渾儀)

The earliest development of the armillary sphere in China goes back to the 1st century BCE.,[364] as they were equipped with a primitive single-ring armillary instrument. This would have allowed them to measure the north polar distance (去極度 , the Chinese form of declination) and measurement that gave the position in a *hsiu* (入宿度 , the Chinese form of right ascension).[365]

During the Western Han Dynasty (202 BC-9 AD), additional developments made by the astronomers Luo Xiahong (落下閎), Xiangyu Wangren, and Geng Shouchang (耿壽昌) advanced the use of the armillary in its early stage of evolution. In 52 BC, it was the astronomer Geng Shou-chang who introduced the fixed equatorial ring to the armillary sphere. In the subsequent Eastern Han Dynasty (23-220 AD) period, the astronomers Fu An and Jia Kui added the elliptical ring by 84 AD. With the famous statesman, astronomer, and inventor Zhang Heng (78-139 AD), the sphere was totally completed in 125 AD, with horizon and meridian rings. It is of great importance to note that the world's first hydraulic (i.e., water-powered) armillary sphere was created by Zhang Heng, who operated his by use of an inflow clepsydra clock (see Zhang's article for more detail).

Abridged armilla (簡儀)

Designed by famous astronomer Guo Shoujing in 1276 AD, it solved most problems found in armillary spheres at that time.

The primary structure of abridged armilla contains two large rings that are perpendicular to each other, of which one is parallel with the equatorial plane and is accordingly called "equatorial ring", and the other is a double ring that is perpendicular to the center of the equatorial ring, revolving around a metallic shaft, and is called "right ascension double ring".

The double ring holds within itself a sighting tube with crosshairs. When observing, astronomers would aim at the star with the sighting tube, whereupon the star's position could be deciphered by observing the dials of the equatorial ring and the right ascension double ring.

A foreign missionary melted the instrument in 1715 AD. The surviving one was built in 1437 AD and was taken to what is now Germany. It was then stored in a French Embassy in 1900, during the Eight-Nation Alliance. Under the pressure of international public discontent, Germany returned the instrument to China. In 1933, it was placed in Purple Mountain Observatory, which prevented it from being destroyed in the Japanese invasion. In the 1980s, it had become seriously eroded and rusted down and was nearly destroyed. In order to restore the device, the Nanjing government spent 11 months to repair it.

Celestial globe (渾象) before Qing Dynasty

Besides star maps, the Chinese also made celestial globes, which show stars' positions like a star map and can present the sky at a specific time. Because of its Chinese name, it is often confused with the armillary sphere, which is just one word different in Chinese (渾象 vs. 渾儀).

According to records, the first celestial globe was made by Geng Shou-chang (耿壽昌) between 70 BC and 50 BC. In the Ming Dynasty, the celestial globe at that time was a huge globe, showing the 28 mansions, celestial equator and ecliptic. None of them have survived.

Celestial globe (天體儀) in the Qing Dynasty

Celestial globes were named 天體儀 ("Miriam celestial bodies") in the Qing Dynasty. The one in Beijing Ancient Observatory was made by Belgian missionary Ferdinand Verbiest (南懷仁) in 1673 AD. Unlike other Chinese celestial globes, it employs 360 degrees rather than the 365.24 degrees (which is a standard in ancient China). It is also the first Chinese globe that shows constellations near to the Celestial South Pole.

Figure 44: *Celestial globe from Qing Dynasty*

The water-powered armillary sphere and celestial globe tower (水運儀象台)

The inventor of the hydraulic-powered armillary sphere was Zhang Heng (78-139 AD) of the Han Dynasty. Zhang was well known for his brilliant applications of mechanical gears, as this was one of his most impressive inventions (alongside his seismograph to detect the cardinal direction of earthquakes that struck hundreds of miles away).

Started by Su Song (蘇頌) and his colleagues in 1086 AD and finished in 1092 AD, his large astronomical clock tower featured an armillary sphere (渾儀), a celestial globe (渾象) and a mechanical chronograph. It was operated by an escapement mechanism and the earliest known chain drive. However, 35 years later, the invading Jurchen army dismantled the tower in 1127 AD upon taking the capital of Kaifeng. The armillary sphere part was brought to Beijing, yet the tower was never successfully reinstated, not even by Su Song's son.

Fortunately, two versions of Su Song's treatise written on his clock tower have survived the ages, so that studying his astronomical clock tower is made possible through medieval texts.

True north and planetary motion

The polymath Chinese scientist Shen Kuo (1031–1095) was not only the first in history to describe the magnetic-needle compass, but also made a more accurate measurement of the distance between the pole star and true north that could be used for navigation. Shen achieved this by making nightly astronomical observations along with his colleague Wei Pu, using Shen's improved design of a wider sighting tube that could be fixed to observe the pole star indefinitely. Along with the pole star, Shen Kuo and Wei Pu also established a project of nightly astronomical observation over a period of five successive years, an intensive work that even would rival the later work of Tycho Brahe in Europe. Shen Kuo and Wei Pu charted the exact coordinates of the planets on a star map for this project and created theories of planetary motion, including retrograde motion.

Foreign influences

Indian astronomy

Buddhism first reached China during the Eastern Han Dynasty, and translation of Indian works on astronomy came to China by the Three Kingdoms era (220–265 CE). However, the most detailed incorporation of Indian astronomy occurred only during the Tang Dynasty (618-907), when a number of Chinese scholars—such as Yi Xing—were versed both in Indian and Chinese astronomy. A system of Indian astronomy was recorded in China as *Jiuzhi-li* (718 CE), the author of which was an Indian by the name of Qutan Xida—a translation of Devanagari Gotama Siddha—the director of the Tang dynasty's national astronomical observatory.[366]

The astronomical table of sines by the Indian astronomer and mathematician Aryabhatan was translated into the Chinese astronomical and mathematical book *Treatise on Astrology of the Kaiyuan Era* (*Kaiyuan Zhanjing*), compiled in 718 AD during the Tang Dynasty. The *Kaiyuan Zhanjing* was compiled by Gautama Siddha, an astronomer and astrologer born in Chang'an, and whose family was originally from India. He was also notable for his translation of the Navagraha calendar into Chinese.

Islamic astronomy in East Asia

Islamic influence on Chinese astronomy was first recorded during the Song dynasty when a Hui Muslim astronomer named Ma Yize introduced the concept of 7 days in a week and made other contributions.[367]

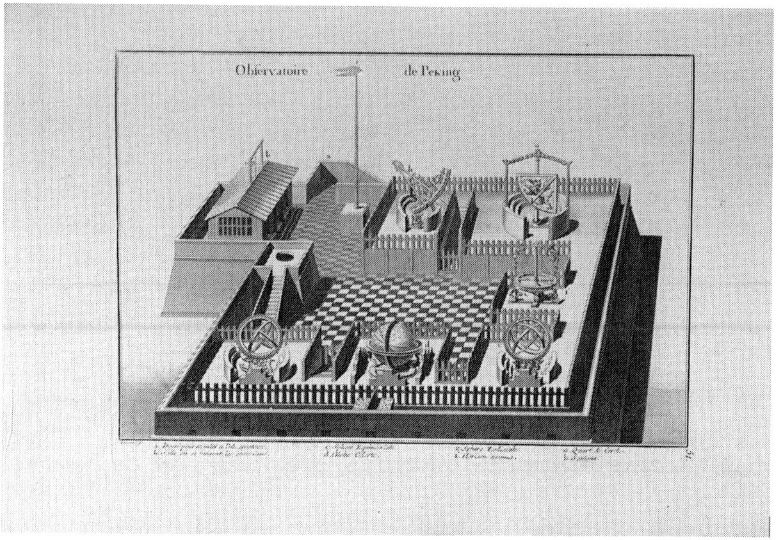

Figure 45: *Early European drawing of the Beijing Ancient Observatory.*

Figure 46: *Gaocheng Astronomical Observatory. It was built in 1276.*

Islamic astronomers were brought to China in order to work on calendar making and astronomy during the Mongol Empire and the succeeding Yuan Dynasty. The Chinese scholar Yelü Chucai accompanied Genghis Khan to Persia in 1210 and studied their calendar for use in the Mongol Empire. Kublai Khan brought Iranians to Beijing to construct an observatory and an institution for astronomical studies.[368]

Several Chinese astronomers worked at the Maragheh observatory, founded by Nasir al-Din al-Tusi in 1259 under the patronage of Hulagu Khan in Persia. One of these Chinese astronomers was Fu Mengchi, or Fu Mezhai.

In 1267, the Persian astronomer Jamal ad-Din, who previously worked at Maragha observatory, presented Kublai Khan with seven Persian astronomical instruments, including a terrestrial globe and an armillary sphere, as well as an astronomical almanac, which was later known in China as the *Wannian Li* ("Ten Thousand Year Calendar" or "Eternal Calendar"). He was known as "Zhama Luding" in China, where, in 1271, he was appointed by Khan as the first director of the Islamic observatory in Beijing, known as the Islamic Astronomical Bureau, which operated alongside the Chinese Astronomical Bureau for four centuries. Islamic astronomy gained a good reputation in China for its theory of planetary latitudes, which did not exist in Chinese astronomy at the time, and for its accurate prediction of eclipses.

Some of the astronomical instruments constructed by the famous Chinese astronomer Guo Shoujing shortly afterwards resemble the style of instrumentation built at Maragheh. In particular, the "simplified instrument" (*jianyi*) and the large gnomon at the Gaocheng Astronomical Observatory show traces of Islamic influence. While formulating the Shoushili calendar in 1281, Shoujing's work in spherical trigonometry may have also been partially influenced by Islamic mathematics, which was largely accepted at Kublai's court.[369] These possible influences include a pseudo-geometrical method for converting between equatorial and ecliptic coordinates, the systematic use of decimals in the underlying parameters, and the application of cubic interpolation in the calculation of the irregularity in the planetary motions.

Emperor Taizu (r. 1368-1398) of the Ming Dynasty (1328–1398), in the first year of his reign (1368), conscripted Han and non-Han astrology specialists from the astronomical institutions in Beijing of the former Mongolian Yuan to Nanjing to become officials of the newly established national observatory.

That year, the Ming government summoned for the first time the astronomical officials to come south from the upper capital of Yuan. There were fourteen of them. In order to enhance accuracy in methods of observation and computation, Emperor Taizu reinforced the adoption of parallel calendar systems, the Han and the Hui. In the following years, the Ming Court appointed several Hui astrologers to hold high positions in the Imperial Observatory. They wrote many books on Islamic astronomy and also manufactured astronomical equipment based on the Islamic system.

The translation of two important works into Chinese was completed in 1383: Zij (1366) and al-Madkhal fi Sina'at Ahkam al-Nujum, Introduction to Astrology (1004).

In 1384, a Chinese astrolabe was made for observing stars based on the instructions for making multi-purposed Islamic equipment. In 1385, the apparatus was installed on a hill in northern Nanjing.

Around 1384, during the Ming Dynasty, Emperor Zhu Yuanzhang ordered the Chinese translation and compilation of Islamic astronomical tables, a task that was carried out by the scholars Mashayihei, a Muslim astronomer, and Wu Bozong, a Chinese scholar-official. These tables came to be known as the *Huihui Lifa* (*Muslim System of Calendrical Astronomy*), which was published in China a number of times until the early 18th century, though the Qing Dynasty had officially abandoned the tradition of Chinese-Islamic astronomy in 1659. The Muslim astronomer Yang Guangxian was known for his attacks on the Jesuit's astronomical sciences.

Jesuit activity in China

The introduction of Western science to China by Jesuit priest astronomers was a mixed blessing during the late sixteenth century and early seventeenth century.

The telescope was introduced to China in the early seventeenth century. The telescope was first mentioned in Chinese writing by Manuel Dias the Younger (Yang Manuo), who wrote his *Tian Wen Lüe* in 1615.[370] In 1626, Johann Adam Schall von Bell (Tang Ruowang) published the Chinese treatise on the telescope known as the *Yuan Jing Shuo* (*The Far-Seeing Optic Glass*).[371] The Chongzhen Emperor (r 1627–1644) of the Ming dynasty acquired the telescope of Johannes Terrentius (or Johann Schreck; Deng Yu-han) in 1634, ten years before the collapse of the Ming Dynasty. However, the impact on Chinese astronomy was limited.

The Jesuit China missions of the sixteenth and seventeenth centuries brought Western astronomy, then undergoing its own revolution, to China and—via João Rodrigues's gifts to Jeong Duwon—to Joseon Korea. After the Galileo affair early in the seventeenth century, the Roman Catholic Jesuit order was required to adhere to geocentrism and ignore the heliocentric teachings of Copernicus and his followers, even though they were becoming standard in European astronomy.[372] Thus, the Jesuits initially shared an Earth-centered and largely pre-Copernican astronomy with their Chinese hosts (i.e., the Ptolemaic-Aristotelian views from Hellenistic times). The Jesuits (such as Giacomo Rho) later introduced Tycho's geoheliocentric model as the standard cosmological model.[373] The Chinese often were fundamentally opposed to this as well, since the Chinese had long believed (from the ancient doctrine of Xuan Ye) that the celestial bodies floated in a void of infinite space. This contradicted the Aristotelian view of solid concentric crystalline spheres, where there was not a void, but a mass of air between the heavenly bodies.

Of course, the views of Copernicus, Galileo, and Tycho Brahe would eventually triumph in European science, and these ideas slowly leaked into China despite Jesuit efforts to curb them in the beginning. In 1627, the Polish Jesuit Michael Boym (Bu Mige) introduced Johannes Kepler's Copernican Rudolphine Tables with much enthusiasm to the Ming court at Beijing. In Adam Schall von Bell's Chinese-written treatise of Western astronomy in 1640, the names of Copernicus (Ge-Bai-Ni), Galileo (Jia-li-lüe), and Tycho Brahe (Di-gu) were formally introduced to China.[374] There were also Jesuits in China who were in favor of the Copernican theory, such as Nicholas Smogulecki and Wenceslaus Kirwitzer. However, Copernican views were not widespread or wholly accepted in China during this time.

Ferdinand Augustin Hallerstein (Liu Songling) created the first spherical astrolabe as the Head of the Imperial Astronomical Bureau from 1739 until 1774. The former Beijing Astronomical observatory, now a museum, still hosts the armillary sphere with rotating rings, which was made under Hallerstein's leadership and is considered the most prominent astronomical instrument.

While in Japan, the Dutch aided the Japanese with the first modern observatory of Japan in 1725, headed by Nakane Genkei, whose observatory of astronomers wholly accepted the Copernican view.[375] In contrast, the Copernican view was not accepted in mainstream China until the early nineteenth century, with the Protestant missionaries such as Joseph Edkins, Alex Wylie, and John Fryer.

Famous Chinese astronomers

- Gan De
- Guo Shoujing
- Shen Kuo
- Shi Shen
- Su Song
- Xu Guangqi
- Zhang Heng

Observatory

Wikimedia Commons has media related to ***Beijing Ancient Observatory***.

- Beijing Ancient Observatory
- Astro Observatory

References

Sources

Book

- Needham, Joseph; Wang Ling (1995) [1959]. *Science and Civilisation in China: Volume 3*. Cambridge, England: Cambridge University Press. ISBN 0-521-05801-5.

Further reading

- *Encyclopaedia of the History of Science, Technology, and Medicine in Non-Western Cultures*, edited by Helaine Selin. Dordrecht: Kluwer, 1997. S.v. "Astronomy in China" by Ho Peng Yoke.
- Sun Xiaochun, "Crossing the Boundaries Between Heaven and Man: Astronomy in Ancient China" in *Astronomy Across Cultures: The History of Non-Western Astronomy*, edited by H. Selin, pp. 423–454. Dordrecht: Kluwer, 2000.
- Chan Ki-hung: *Chinese Ancient Star Map*, Leisure and Cultural Services Department, 2002, ISBN 962-7054-09-7
- *Gems of the ancient Chinese astronomy relics*, ISBN 962-7797-03-0

External links

- The Mathematics of the Chinese Calendar[376] by Helmar Aslaksen
- Boundaries Crossing: Western Astronomy in Confucian China, 1600-1800[377] by Pingyi Chu
- Homepage of the National Astronomical Observatories, Chinese Academy of Sciences[378]
- Chinese astronomy at the University of Maine[379]

Maya calendar

Maya civilization

- People
- Society
- Languages
- Writing
- Religion
- Mythology
- Sacrifice
- Cities
- Architecture
- Astronomy
- Calendar
- Stelae
- Art
- Textiles
- Trade
- Music
- Dance
- Medicine
- Cuisine

History

Preclassic Maya

Classic Maya collapse

Spanish conquest of the Maya

- Yucatán
- Chiapas
- Guatemala
- Petén

- v
- t
- e[380]

The **Maya calendar** is a system of calendars used in pre-Columbian Mesoamerica and in many modern communities in the Guatemalan highlands,[381] Veracruz, Oaxaca and Chiapas, Mexico.[382]

The essentials of the Maya calendar are based upon a system which had been in common use throughout the region, dating back to at least the 5th century BCE. It shares many aspects with calendars employed by other earlier Mesoamerican civilizations, such as the Zapotec and Olmec and contemporary or later ones such as the Mixtec and Aztec calendars.

By the Maya mythological tradition, as documented in Colonial Yucatec accounts and reconstructed from Late Classic and Postclassic inscriptions, the deity Itzamna is frequently credited with bringing the knowledge of the calendar system to the ancestral Maya, along with writing in general and other foundational aspects of Maya culture.[383]

Overview

The Maya calendar consists of several cycles or *counts* of different lengths. The 260-day count is known to scholars as the *Tzolkin*, or *Tzolk'in*.[384] The Tzolkin was combined with a 365-day vague solar year known as the Haab' to form a synchronized cycle lasting for 52 Haab', called the *Calendar Round*. The Calendar Round is still in use by many groups in the Guatemalan highlands.[385]

A different calendar was used to track longer periods of time and for the inscription of calendar dates (i.e., identifying when one event occurred in relation to others). This is the *Long Count*. It is a count of days since a mythological starting-point.[386] According to the correlation between the Long Count and Western calendars accepted by the great majority of Maya researchers (known as the Goodman-Martinez-Thompson, or GMT, correlation), this starting-point is equivalent to August 11, 3114 BCE in the proleptic Gregorian calendar or September 6, in the Julian calendar (–3113 astronomical). The GMT correlation was chosen by John Eric Sydney Thompson in 1935 on the basis of earlier correlations by Joseph Goodman in 1905 (August 11), Juan Martínez Hernández in 1926 (August 12) and Thompson himself in 1927 (August 13).[387] By its linear nature, the Long Count was capable of being extended to refer to any date far into the past or future. This calendar involved the use of a positional notation system, in which each position signified an increasing multiple of the number of days. The Maya numeral system was essentially vigesimal (i.e., base-20) and each unit of a given position represented 20 times the unit of the position which preceded it. An important exception was made for the second-order place value, which instead represented 18×20, or 360 days, more closely approximating the solar year than would $20 \times 20 = 400$ days. It should be noted however that the cycles of the Long Count are independent of the solar year.

Many Maya Long Count inscriptions contain a supplementary series, which provides information on the lunar phase, number of the current lunation in a series of six and which of the nine Lords of the Night rules.

Less-prevalent or poorly understood cycles, combinations and calendar progressions were also tracked. An *819-day Count* is attested in a few inscriptions. Repeating sets of 9 days (see below "Nine lords of the night")[388] associated with different groups of deities, animals and other significant concepts are also known.

Tzolk'in

The *tzolk'in* (in modern Maya orthography; also commonly written *tzolkin*) is the name commonly employed by Mayanist researchers for the Maya Sacred Round or 260-day calendar. The word *tzolk'in* is a neologism coined in Yucatec Maya, to mean "count of days" (Coe 1992). The various names of this calendar as used by precolumbian Maya people are still debated by scholars. The Aztec calendar equivalent was called *Tonalpohualli*, in the Nahuatl language.

The tzolk'in calendar combines twenty day names with the thirteen day numbers to produce 260 unique days. It is used to determine the time of religious and ceremonial events and for divination. Each successive day is numbered from 1 up to 13 and then starting again at 1. Separately from this, every day is given a name in sequence from a list of 20 day names:

Tzolk'in calendar: named days and associated glyphs

Seq. Num. [1]	Day Name [2]	Glyph example [3]	16th-c. Yucatec [4]	reconstructed Classic Maya [5]	Seq. Num. [1]	Day Name [2]	Glyph example [3]	16th-c. Yucatec [4]	reconstructed Classic Maya [5]
01	**Imix'**		Imix	Imix (?) / Ha' (?)	11	**Chuwen**		Chuen	(unknown)
02	**Ik'**		Ik	Ik'	12	**Eb'**		Eb	(unknown)
03	**Ak'b'al**		Akbal	Ak'b'al (?)	13	**B'en**		Ben	C'klab
04	**K'an**		Kan	K'an (?)	14	**Ix**		Ix	Hix (?)
05	**Chikchan**		Chicchan	(unknown)	15	**Men**		Men	(unknown)
06	**Kimi**		Cimi	Cham (?)	16	**K'ib'**		Cib	(unknown)
07	**Manik'**		Manik	Manich' (?)	17	**Kab'an**		Caban	Chab' (?)
08	**Lamat**		Lamat	Ek' (?)	18	**Etz'nab'**		Etznab	(unknown)

09	**Muluk**		Muluc	(unknown)	**19**	**Kawak**		Cauac	(unknown)
10	**Ok**		Oc	(unknown)	**20**	**Ajaw**		Ahau	Ajaw

NOTES:
1. The sequence number of the named day in the Tzolk'in calendar
2. Day name, in the standardized and revised orthography of the Guatemalan Academia de Lenguas Mayas[389]
3. An example glyph (logogram) for the named day. Note that for most of these several different forms are recorded; the ones shown here are typical of carved monumental inscriptions (these are "cartouche" versions)
4. Day name, as recorded from 16th-century Yukatek Maya accounts, principally Diego de Landa; this orthography has (until recently) been widely used
5. In most cases, the actual day name as spoken in the time of the Classic Period (c. 200–900) when most inscriptions were made is not known. The versions given here (in Classic Maya, the main language of the inscriptions) are reconstructed on the basis of phonological evidence, if available; a '?' symbol indicates the reconstruction is tentative.[390]

Some systems started the count with 1 Imix', followed by 2 Ik', 3 Ak'b'al, etc. up to 13 B'en. The day numbers then start again at 1 while the named-day sequence continues onwards, so the next days in the sequence are 1 Ix, 2 Men, 3 K'ib', 4 Kab'an, 5 Etz'nab', 6 Kawak and 7 Ajaw. With all twenty named days used, these now began to repeat the cycle while the number sequence continues, so the next day after 7 Ajaw is 8 Imix'. The repetition of these interlocking 13- and 20-day cycles therefore takes 260 days to complete (that is, for every possible combination of number/named day to occur once).

The earliest known inscription with a Tzolk'in is an Olmec earspool with 2 Ahau 3 Ceh - 6.3.10.9.0, September 2, -678 (Julian astronomical).

Haab'

Haab' months: names and glyphs[391] in sequence

Seq. Num.	Yucatec name	Hiero-glyph	Meaning of glyph[392]	Seq. Num.	Yucatec name	Hiero-glyph	Meaning of glyph
1	Pop	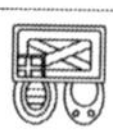		10	Yax		green
2	Wo'			11	Sak'	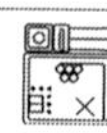	white
3	Sip			12	Keh		red
4	Sotz'			13	Mak		
5	Sek	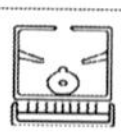		14	K'ank'in	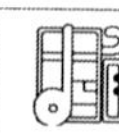	
6	Xul			15	Muwan'		
7	Yaxk'in'	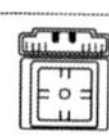		16	Pax	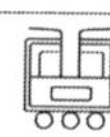	
8	Mol			17	K'ayab		
9	Ch'en	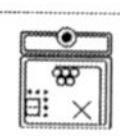	black	18	Kumk'u	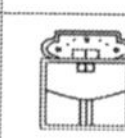	
				19	Wayeb'	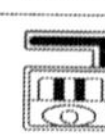	five unlucky days

The Haab' was made up of eighteen months of twenty days each plus a period of five days ("nameless days") at the end of the year known as *Wayeb'* (or *Uayeb* in 16th-century orthography). The five days of Wayeb' were thought to be a dangerous time. Foster (2002) writes, "During Wayeb, portals between the mortal realm and the Underworld dissolved. No boundaries prevented the ill-intending deities from causing disasters." To ward off these evil spirits, the Maya had customs and rituals they practiced during Wayeb'. For example, people avoided leaving their houses and washing or combing their hair. Bricker (1982) estimates that the Haab' was first used around 550 BCE with a starting point of the winter solstice.[393]

The Haab' month names are known today by their corresponding names in colonial-era Yukatek Maya, as transcribed by 16th-century sources (in particular, Diego de Landa and books such as the *Chilam Balam* of Chumayel). Phonemic analyses of Haab' glyph names in pre-Columbian Maya inscriptions have demonstrated that the names for these twenty-day periods varied considerably from region to region and from period to period, reflecting differences in the base language(s) and usage in the Classic and Postclassic eras predating their recording by Spanish sources.[394]

Each day in the Haab' calendar was identified by a day number in the month followed by the name of the month. Day numbers began with a glyph translated as the "seating of" a named month, which is usually regarded as day 0 of that month, although a minority treat it as day 20 of the month preceding the named month. In the latter case, the seating of Pop is day 5 of Wayeb'. For the majority, the first day of the year was 0 Pop (the seating of Pop). This was followed by 1 Pop, 2 Pop as far as 19 Pop then 0 Wo, 1 Wo and so on.

The Haab' year had exactly 365 days and ignored the extra quarter of day (approximately) in the actual tropical year. This meant that the seasons moved with respect to the calendar year by a quarter of a day each year, so that the calendar months named after particular seasons no longer corresponded to these seasons after a few centuries.

Calendar Round

A Calendar Round date is a date that gives both the Tzolk'in and Haab'. This date will repeat after 52 Haab' years or 18,980 days, a Calendar Round. For example, the current creation started on 4 Ahau 8 Kumk'u. When this date recurs it is known as a Calendar Round completion.

Arithmetically, the duration of the Calendar Round is the least common multiple of 260 and 365; 18,980 is 73×260 Tzolk'in days and 52×365 Haab' days.[395]

Not every possible combination of Tzolk'in and Haab' can occur. For Tzolk'in days Imix, Kimi, Chwen and Kib', the Haab' day can only be 4, 9, 14 or 19; for Ik', Manik', Eb' and Kab'an, the Haab' day can only be 0, 5, 10 or 15; for Akb'al', Lamat, B'en and Etz'nab', the Haab' day can only be 1, 6, 11 or 16; for K'an, Muluk, Ix and Kawak, the Haab' day can only be 2, 7, 12 or 17; and for Chikchan, Ok, Men and Ajaw, the Haab' day can only be 3, 8, 13 or 18.

Year Bearer

A "Year Bearer" is a Tzolk'in day name that occurs on the first day of the Haab'. If the first day of the Haab' is 0 Pop, then each 0 Pop will coincide with a Tzolk'in date, for example, 1 Ik' 0 Pop. Since there are twenty Tzolk'in day names and the Haab' year has 365 days (20*18 + 5), the Tzolk'in name for each succeeding Haab' zero day will be incremented by 5 in the cycle of day names like this:

1 Ik' 0 Pop
2 Manik' 0 Pop
3 Eb' 0 Pop
4 Kab'an 0 Pop
5 Ik' 0 Pop...

Only these four of the Tzolk'in day names can coincide with 0 Pop and these four are called the "Year Bearers".

"Year Bearer" literally translates a Mayan concept.[396] Its importance resides in two facts. For one, the four years headed by the Year Bearers are named after them and share their characteristics; therefore, they also have their own prognostications and patron deities.[397] Moreover, since the Year Bearers are geographically identified with boundary markers or mountains, they help define the local community.[398]

The classic system of Year Bearers described above is found at Tikal and in the Dresden Codex. During the Late Classic period a different set of Year Bearers was in use in Campeche. In this system, the Year Bearers were the Tzolk'in that coincided with 1 Pop. These were Ak'b'al, Lamat, B'en and Edz'nab. During the Post-Classic period in Yucatán a third system was in use. In this system the Year Bearers were the days that coincided with 2 Pop: K'an, Muluc, Ix and Kawak. This system is found in the Chronicle of Oxkutzcab. In addition, just before the Spanish conquest in Mayapan the Maya began to number the days of the Haab' from 1 to 20. In this system the Year Bearers are the same as in the 1 Pop – Campeche system. The Classic Year Bearer system is still in use in the Guatemalan highlands[399] and in Veracruz, Oaxaca and Chiapas, Mexico.[400]

Figure 47: *East side of stela C, Quirigua with the mythical creation date of 13 baktuns, 0 katuns, 0 tuns, 0 uinals, 0 kins, 4 Ahau 8 Cumku – August 11, 3114 BCE in the proleptic Gregorian calendar.*

Long Count

Since Calendar Round dates repeat every 18,980 days, approximately 52 solar years, the cycle repeats roughly once each lifetime, so a more refined method of dating was needed if history was to be recorded accurately. To specify dates over periods longer than 52 years, Mesoamericans used the Long Count calendar.

The Maya name for a day was *k'in*. Twenty of these k'ins are known as a *winal* or *uinal*. Eighteen winals make one *tun*. Twenty tuns are known as a *k'atun*. Twenty k'atuns make a *b'ak'tun*.

The Long Count calendar identifies a date by counting the number of days from the Mayan creation date 4 Ahaw, 8 Kumk'u (August 11, 3114 BC in the proleptic Gregorian calendar or September 6 in the Julian calendar -3113 astronomical dating). But instead of using a base-10 (decimal) scheme like Western numbering, the Long Count days were tallied in a modified base-20 scheme. Thus 0.0.0.1.5 is equal to 25 and 0.0.0.2.0 is equal to 40. As the winal unit resets after only counting to 18, the Long Count consistently uses base-20 only if the tun is considered the primary unit of measurement, not the k'in; with the k'in and winal units being the number of days in the tun. The Long

Count 0.0.1.0.0 represents 360 days, rather than the 400 in a purely base-20 (vigesimal) count.

There are also four rarely used higher-order cycles: piktun, kalabtun, k'inchiltun and alautun.

Since the Long Count dates are unambiguous, the Long Count was particularly well suited to use on monuments. The monumental inscriptions would not only include the 5 digits of the Long Count, but would also include the two tzolk'in characters followed by the two haab' characters.

Misinterpretation of the Mesoamerican Long Count calendar was the basis for a popular belief that a cataclysm would take place on December 21, 2012. December 21, 2012 was simply the day that the calendar went to the next b'ak'tun, at Long Count 13.0.0.0.0. The date on which the calendar will go to the next piktun (a complete series of 20 b'ak'tuns), at Long Count 1.0.0.0.0.0, will be on October 13, 4772.

Sandra Noble, executive director of the Mesoamerican research organization Foundation for the Advancement of Mesoamerican Studies, Inc. (FAMSI), notes that "for the ancient Maya, it was a huge celebration to make it to the end of a whole cycle". She considers the portrayal of December 2012 as a doomsday or cosmic-shift event to be "a complete fabrication and a chance for a lot of people to cash in."[401]

Table of Long Count units

Long Count unit	Long Count period	Days	Approximate Solar Years
1 K'in		1	
1 Winal	20 K'in	20	
1 Tun	18 Winal	360	1
1 K'atun	20 Tun	7,200	20
1 B'ak'tun	20 K'atun	144,000	394
1 Piktun	20 B'ak'tun	2,880,000	7,885
1 Kalabtun	20 Piktun	57,600,000	157,704
1 K'inchiltun	20 Kalabtun	1,152,000,000	3,154,071
1 Alautun	20 K'inchiltun	23,040,000,000	63,081,429

Supplementary Series

Many Classic period inscriptions include a series of glyphs known as the Supplementary Series. The operation of this series was largely worked out by John E. Teeple (1874–1931). The Supplementary Series most commonly consists of the following elements:

Lords of the Night

Each night was ruled by one of the nine lords of the underworld. This nine-day cycle was usually written as two glyphs: a glyph that referred to the Nine Lords as a group, followed by a glyph for the lord that would rule the next night.

Lunar Series

A lunar series generally is written as five glyphs that provide information about the current lunation, the number of the lunation in a series of six, the current ruling lunar deity and the length of the current lunation.

Moon age

The Maya counted the number of days in the current lunation. They used two systems for the zero date of the lunar cycle: either the first night they could see the thin crescent moon or the first morning when they could not see the waning moon.[402] The age of the moon was depicted by a set of glyphs that mayanists coined glyphs D and E:

- A new moon glyph was used for day zero in the lunar cycle.
- D glyphs were used for lunar ages for days 1 through 19, with the number of days that had passed from the new moon.
- For lunar ages 20 to 30, an E glyph was used, with the number of days from 20.

Count of Lunations

The Maya counted the lunations. This cycle appears in the lunar series as two glyphs that modern scholars call the 'C' and 'X' glyphs. The C glyph could be prefixed with a number indicating the lunation. No prefixing number meant one, whereas the numbers two through six indicated the other lunations.[403,404] There was also a part of the C glyph that indicated where this fell in a larger cycle of 18 lunations. Accompanying the C glyph was the 'X' glyph that showed a similar pattern of 18 lunations.[405,406]

Lunation length

The present era lunar synodic period is about 29.5305877 mean solar days or about 29 days 12 hours 44 minutes and $2+{}^{7}/_{9}$ seconds. As a whole number, the number of days per lunation will be either 29 or 30 days, with the 30-day intervals necessarily occurring slightly more frequently than the 29-day intervals. The Maya wrote whether the lunar month was 29 or 30 days as two glyphs: a glyph for lunation length followed by either a glyph made up of a moon glyph over a bundle with a suffix of 9 for a 29-day lunation or a moon glyph with a suffix of 10 for a 30-day lunation. Since the Maya didn't use fractions, lunations were approximated by using the formula that there were 149 lunations completed in 4400 days, which yielded a rather short mean month of exactly ${}^{4400}/_{149} = 29+{}^{79}/_{149}$ days = 29 days 12 hours 43 minutes and $29+{}^{59}/_{149}$ seconds, or about 29.5302 days.[407]

819-day count

Some Mayan monuments include glyphs that record an 819-day count in their Initial Series. These can also be found in the Dresden codex.[408] This is described in Thompson.[409] More examples of this can be found in Kelley.[410] Each group of 819 days was associated with one of four colors and the cardinal direction with which it was associated — black corresponded to west, red to east, white to north and yellow to south.

The 819-day count can be described several ways: Most of these are referred to using a "Y" glyph and a number. Many also have a glyph for K'awill — the god with a smoking mirror in his head. K'awill has been suggested as having a link to Jupiter.[411] In the Dresden codex almanac 59 there are Chaacs of the four colors. The accompanying texts begin with a directional glyph and a verb for 819-day-count phrases. Anderson provides a detailed description of the 819-day count.

Short Count

During the late Classic period the Maya began to use an abbreviated short count instead of the Long Count. An example of this can be found on altar 14 at Tikal.[412] In the kingdoms of Postclassic Yucatán, the Short Count was used instead of the Long Count. The cyclical Short Count is a count of 13 k'atuns (or 260 tuns), in which each k'atun was named after its concluding day, Ahau ('Lord'). 1 Imix was selected as the recurrent 'first day' of the cycle, corresponding to 1 Cipactli in the Aztec day count. The cycle was counted from katun 11 Ahau to katun 13 Ahau, with the coefficients of the katuns' concluding days running in the order 11 – 9 – 7 – 5 – 3 – 1 – 12 – 10 – 8 –

6 – 4 – 2 – 13 Ahau (since a division of 20 × 360 days by 13 falls 2 days short). The concluding day 13 Ahau was followed by the re-entering first day 1 Imix. This is the system as found in the colonial Books of Chilam Balam. In characteristic Mesoamerican fashion, these books project the cycle onto the landscape, with 13 *Ahauob* 'Lordships' dividing the land of Yucatán into 13 'kingdoms'.[413]

References

Aveni, Anthony F. (2001). *Skywatchers* (originally published as: *Skywatchers of Ancient Mexico* [1980], revised and updated ed.). Austin: University of Texas Press. ISBN 0-292-70504-2. OCLC 45195586[414].

Boot, Erik (2002). *A Preliminary Classic Maya-English/English-Classic Maya Vocabulary of Hieroglyphic Readings*[415] (PDF). Mesoweb. Retrieved 2006-11-10.

Bricker, Victoria R. (February 1982). "The Origin of the Maya Solar Calendar". *Current Anthropology*. Chicago, IL: University of Chicago Press, sponsored by Wenner-Gren Foundation for Anthropological Research. **23** (1): 101–103. ISSN 0011-3204[416]. OCLC 62217742[417]. doi: 10.1086/202782[418].

Coe, Michael D. (1965). "A Model of Ancient Maya Community Structure in the Maya Lowlands". *Southwestern Journal of Anthropology*. **21**.

Coe, Michael D. (1987). *The Maya* (4th revised ed.). London and New York: Thames & Hudson. ISBN 0-500-27455-X. OCLC 15895415[419].

Coe, Michael D. (1992). *Breaking the Maya Code*. London: Thames & Hudson. ISBN 0-500-05061-9. OCLC 26605966[420].

Foster, Lynn V. (2002). *Handbook to Life in the Ancient Maya World*. with Foreword by Peter Mathews. New York: Facts on File. ISBN 0-8160-4148-2. OCLC 50676955[421].

Ivanoff, Pierre (1971). *Mayan Enigma: The Search for a Lost Civilization*. Elaine P. Halperin (trans.) (translation of *Découvertes chez les Mayas*, English ed.). New York: Delacorte Press. ISBN 0-440-05528-8. OCLC 150172[422].

Jones, Christopher (1984). *Deciphering Maya Hieroglyphs*. Carl P. Beetz (illus.) (prepared for Weekend Workshop April 7 and 8, 1984, 2nd ed.). Philadelphia: University Museum, University of Pennsylvania. OCLC 11641566[423].

Kettunen, Harri; Christophe Helmke (2005). *Introduction to Maya Hieroglyphs: 10th European Maya Conference Workshop Handbook*[424] (PDF).

Leiden, Netherlands: *Wayeb* and Leiden University. Retrieved 2006-06-08.

Linden, John H. (1996). *The Deity Head Variants of the C Glyph*. The Eight Palenque Round Table, 1993. pp. 343–356.

MacDonald, G. Jeffrey (27 March 2007). "Does Maya calendar predict 2012 apocalypse?"[425]. *USA Today*. McLean, VA: Gannett Company. ISSN 0734-7456[426]. Archived from the original[427] on 2008-03-16. Retrieved 2009-05-28.

Milbrath, Susan (1999). *Star Gods of the Maya: Astronomy in Art, Folklore, and Calendars*. The Linda Schele series in Maya and pre-Columbian studies. Austin: University of Texas Press. ISBN 0-292-75225-3. OCLC 40848420[428].

Miller, Mary; Karl Taube (1993). *The Gods and Symbols of Ancient Mexico and the Maya: An Illustrated Dictionary of Mesoamerican Religion*. London: Thames and Hudson. ISBN 0-500-05068-6. OCLC 27667317[429].

Rice, Prudence M., *Maya Calendar Origins: Monuments, Mythistory, and the Materialization of Time* (Austin, TX: University of Texas Press, 2007) .

Robinson, Andrew (2000). *The Story of Writing: Alphabets, Hieroglyphs and Pictograms*. London and New York: Thames & Hudson. ISBN 0-500-28156-4. OCLC 59432784[430].

Roys, Ralph L. (1967). *The Book of Chilam Balam of Chumayel*. Norman: University of Oklahoma Press.

Schele, Linda; David Freidel (1992). *A Forest of Kings: The Untold Story of the Ancient Maya* (originally published New York: Morrow © 1990, pbk reprint ed.). New York: Harper Perennial. ISBN 0-688-11204-8. OCLC 145324300[431].

Schele, Linda; Nickolai Grube; Federico Fahsen (October 1992). "The Lunar Series in Classic Maya Inscriptions: New Observation and Interpretations". *Texas Notes on Precolumbian Art, Writing, and Culture* (29).

Tedlock, Barbara (1992). *Time and the Highland Maya* (rev. ed.). Albuquerque: University of New Mexico Press. ISBN 0-8263-0577-6. OCLC 7653289[432].

Teeple, John E. (November 1931). "Maya Astronomy"[433] (PDF). *Contributions to American Archaeology*. Volume I (Pub. 403 ed.). Washington D.C.: Carnegie Institution of Washington. pp. 29–116.

Tedlock, Dennis notes, trans., eds. (1985). *Popol Vuh: The Definitive Edition of the Mayan Book of the Dawn of Life and the Glories of Gods and Kings*. with commentary based on the ancient knowledge of the modern Quiché Maya. New York: Simon & Schuster. ISBN 0-671-45241-X. OCLC 11467786[434].

Thomas, Cyrus (1897). "Day Symbols of the Maya Year"[435]. In J. W. Powell. *Sixteenth Annual Report of the Bureau of American Ethnology to the Secretary of the Smithsonian Institution, 1894–1895* (EBook online reproduction). Washington DC: Bureau of American Ethnology, Smithsonian Institution; U.S. Government Printing Office. pp. 199–266. OCLC 14963920[436] – via Project Gutenberg.Wikipedia:Link rot

Thompson, J. Eric S. (1960). *Maya Hieroglyphic Writing: An Introduction*. Civilization of the American Indian Series, No. 56 (3rd ed.). Norman: University of Oklahoma Press. ISBN 0-8061-0447-3. OCLC 275252[437].

Tozzer, Alfred M. notes, trans., eds. (1941). *Landa's Relación de las cosas de Yucatán: a translation*. Papers of the Peabody Museum of American Archaeology and Ethnology, Harvard University vol. 18. Charles P. Bowditch and Ralph L. Roys (additional trans.) (translation of Diego de Landa's *Relación de las cosas de Yucatán* [original c. 1566], with notes, commentary, and appendices incorporating translated excerpts of works by Gaspar Antonio Chi, Tomás López Medel, Francisco Cervantes de Salazar, and Antonio de Herrera y Tordesillas. English ed.). Cambridge, MA: Peabody Museum of Archaeology and Ethnology. OCLC 625693[438].

Voss, Alexander (2006). "Astronomy and Mathematics". In Nikolai Grube. *Maya: Divine Kings of the Rain Forest*. Eva Eggebrecht and Matthias Seidel (assistant eds.). Cologne, Germany: Könemann. pp. 130–143. ISBN 978-3-8331-1957-6. OCLC 71165439[439].

External links

Wikimedia Commons has media related to ***Maya calendar***.

- *Day Symbols of the Maya Year* at Project Gutenberg 1897 text by Cyrus Thomas
- date converter at FAMSI[440] This converter uses the Julian/Gregorian calendar and includes the 819 day cycle and lunar age.
- Interactive Maya Calendars[441]

- Maya Calendar, Date conversions, contemporary year version, Tzolkin and Haab day in Calendar Rounds[442]

Aztec calendar

Aztec Empire
Mythology
Military · Codices
History
Spanish conquest of Mexico
La Noche Triste
Engineering
Education
Religion
Cuisine
Architecture

- v
- t
- e[443]

The **Aztec or Mexica calendar** is the calendar system that was used by the Aztecs as well as other Pre-Columbian peoples of central Mexico. It is one of the Mesoamerican calendars, sharing the basic structure of calendars from throughout ancient Mesoamerica.

The calendar consisted of a 365-day calendar cycle called *xiuhpohualli* (year count) and a 260-day ritual cycle called *tonalpohualli* (day count). These two cycles together formed a 52-year "century," sometimes called the "calendar round". The *xiuhpohualli* is considered to be the agricultural calendar, since it is based on the sun, and the *tonalpohualli* is considered to be the sacred calendar.

Figure 48: *The Aztec Aztec calendar stone, also called the Sun Stone, on display at the National Museum of Anthropology, Mexico City.*

Tonalpohualli

The *tonalpohualli* ("day count") consists of a cycle of 260 days, each day signified by a combination of a number from 1 to 13, and one of the twenty day signs. With each new day, both the number and day sign would be incremented: 1 Crocodile is followed by 2 Wind, 3 House, 4 Lizard, and so forth up to 13 Reed, after which the cycle of numbers would restart (though the twenty day signs had not yet been exhausted) resulting in 1 Jaguar, 2 Eagle, and so on, as the days immediately following 13 Reed. This cycle of number and day signs would continue similarly until the 20th week, which would start on 1 Rabbit, and end on 13 Flower. It would take a full 260 days (13×20) for the two cycles (of twenty day signs, and thirteen numbers) to realign and repeat the sequence back on 1 Crocodile.

Day signs

The set of day signs used in central Mexico is identical to that used by Mixtecs, and to a lesser degree similar to those of other Mesoamerican calendars. Each of the day signs also bears an association with one of the four cardinal directions.Wikipedia:Verifiability

There is some variation in the way the day signs were drawn or carved. Those here were taken from the Codex Magliabechiano.

Image	Nahuatl name	Pronunciation	English translation	Direction
	*Cipactli*	[siˈpáktɬi]	Crocodile Alligator Caiman Crocodilian Monster Dragon	East
	*Ehēcatl*	[eʔˈé:katɬ]	Wind	North
	*Calli*	[ˈkáɬ:i]	House	West
	*Cuetzpalin*	[kʷetsˈpálin̩]	Lizard	South
	*Cōātl*	[ˈkú˕:wa:tɬ]	Serpent Snake	East
	*Miquiztli*	[miˈkístɬi]	Death	North
	*Mazātl*	[ˈmása:tɬ]	Deer Animal	West
	*Tōchtli*	[ˈtú˕:tʃtɬi]	Rabbit	South
	*Ātl*	[ˈa:tɬ]	Water	East
	*Itzcuīntli*	[itsˈkʷí:n̩tɬi]	Dog	North

Image	Nahuatl name	Pronunciation	English translation	Direction
	*Ozomatli* *Ozomahtli*	[u˕su˕ˈmáʔtɬi]	Monkey	West
	*Malīnalli*	[mali:ˈnáɬ:i]	Grass	South
	*Ācatl*	[ˈá:katɬ]	Reed	East

	*Ōcēlōtl*	[u˕:ˈsé:lu˕:tɬ]	Jaguar	North
	*Cuāuhtli*	[ˈkʷá:ʍtɬi]	Eagle	West
	*Cōzcacuāuhtli*	[ku˕:skaˈkʷá:ʍtɬi]	Vulture	South
	*Olīn*	[ˈú˕li:n̥]	Movement Quake Earthquake	East
	*Tecpatl*	[ˈtékpatɬ]	Flint Flint Knife	North
	*Quiyahuitl*	[kiˈjáwitɬ]	Rain	West
	*Xōchitl*	[ˈʃú˕:tʃitɬ]	Flower	South

Wind and Rain are represented by images of their associated gods, Ehecatl and Tlaloc (respectively).

Other marks on the stone showed the current world and also the worlds before this one. Each world was called a sun, and each sun had its own species of inhabitants. The Aztecs believed that they were in the fifth sun and like all of the suns before them they would also eventually perish due to their own imperfections. Every fifty two years was marked out because they believed that fifty two years was a life cycle and at the end of any given life cycle the gods could take away all that they have and destroy the world.

Trecenas

The 260 days of the sacred calendar were grouped into twenty periods of thirteen days each. Scholars usually refer to these thirteen-day "weeks" as *trecenas*, using a Spanish term derived from *trece* "thirteen" (just as the Spanish term *docena* "dozen" is derived from *doce* "twelve"). The original Nahuatl term is not known.

Each *trecena* is named according to the calendar date of the first day of the thirteen days in that *trecena*. In addition, each of the twenty *trecenas* in the 260-day cycle had its own tutelary deity:

Trecena	Deity	Trecena	Deity
1 Crocodile	Ometeotl	1 Monkey	Patecatl
1 Jaguar	Quetzalcoatl	1 Lizard	Itztlacoliuhqui
1 Deer	Tepeyollotl	1 Quake	Tlazolteotl
1 Flower	Huehuecoyotl	1 Dog	Xipe Totec
1 Reed	Chalchiuhtlicue	1 House	Itzpapalotl
1 Death	Tonatiuh	1 Vulture	Xolotl
1 Rain	Tlaloc	1 Water	Chalchiuhtotolin
1 Grass	Mayahuel	1 Wind	Chantico
1 Snake	Xiuhtecuhtli	1 Eagle	Xochiquetzal
1 Flint	Mictlantecuhtli	1 Rabbit	Xiuhtecuhtli

Xiuhpohualli

Veintena (twenty); metzli (moon)

"In ancient times the year was composed of eighteen months, and thus it was observed by the native people. Since their months were made of no more than twenty days, these were all the days contained in a month, because they were not guided by the moon but by the days; therefore, the year had eighteen months. The days of the year were counted twenty by twenty." Diego Durán

Xiuhpohualli is the Aztec year (*xihuitl*) count (*pohualli*). One year consists of 360 named days and 5 nameless (*nemontemi*). These 'extra' days are thought to be unlucky. The year was broken into 18 periods of twenty days each, sometimes compared to the Julian month. The Aztec word for moon is *metztli* but whatever name was used for these periods is unknown. Through Spanish usage, the 20-day period of the Aztec calendar has become commonly known as a *veintena*.

Each 20-day period started on *Cipactli* (Crocodile) for which a festival was held. The eighteen *veintena* are listed below. The dates are from early eye-witnesses. Each wrote what they saw. Bernardino de Sahagún's date precedes the observations of Diego Durán by several decades and is believed to be more recent to the surrender. Both are shown to emphasize the fact that the beginning of the Native new year became non-uniform as a result of an absence of the unifying force of Tenochtitlan after the Mexica defeat.

Duran Time	Sahagun Time	Fiesta Names	Symbol	English Translation
1. MAR 01 - MAR 20	1. FEB 02 - FEB 21	*Atlcahualo, Cuauhitlehua*		Ceasing of Water, Rising Trees
2. MAR 21 - APR 09	2. FEB 22 - MAR 13	*Tlacaxipehualiztli*		Rites of Fertility; Xipe-Totec ("the flayed one")
3. APR 10 - APR 29	3. MAR 14 - APR 02	*Tozoztontli*		Lesser Perforation
4. APR 30 - MAY 19	4. APR 03 - APR 22	*Huey Tozoztli*		Greater Perforation
5. MAY 20 - JUN 08	5. APR 23 - MAY 12	*Toxcatl*		Dryness
6. JUN 09 - JUN 28	6. MAY 13 - JUN 01	*Etzalcualiztli*		Eating Maize and Beans
7. JUN 29 - JULY 18	7. JUN 02 - JUN 21	*Tecuilhuitontli*		Lesser Feast for the Revered Ones
8. JULY 19 - AUG 07	8. JUN 22 - JUL 11	*Huey Tecuilhuitl*		Greater Feast for the Revered Ones
9. AUG 08 - AUG 27	9. JUL 12 - JUL 31	*Tlaxochimaco, Miccailhuitontli*		Bestowal or Birth of Flowers, Feast to the Revered Deceased
10. AUG 28 - SEP 16	10. AUG01 - AUG 20	*Xócotl huetzi, Huey Miccailhuitl*		Feast to the Greatly Revered Deceased
11. SEPT 17 - OCT 06	11. AUG 21 - SEPT 09	*Ochpaniztli*		Sweeping and Cleaning
12. OCT 07 - OCT 26	12. SEPT10 - SEPT 29	*Teotleco*		Return of the Gods
13. OCT 27 - NOV 15	13. SEPT 30 - OCT 19	*Tepeilhuitl*		Feast for the Mountains
14. NOV 16 - DEC 05	14. OCT 20 - NOV 8	*Quecholli*		Precious Feather
15. DEC 06 - DEC 25	15. NOV 09 - NOV 28	*Panquetzaliztli*		Raising the Banners
16. DEC 26 - JAN 14	16. NOV 29 - DEC 18	*Atemoztli*		Descent of the Water
17. JAN 15 - FEB 03	17. DEC 19 - JAN 07	*Tititl*		Stretching for Growth

18. FEB 04 - FEB 23	18. JAN 08 - JAN 27	*Izcalli*		Encouragement for the Land & People
18u. FEB 24 - FEB 28	18u.JAN 28 - FEB 01	*nemontemi* (5 day period)		Empty days (no specific activities or holidays)

Reconstruction of the Solar Calendar

For many centuries scholars had tried to reconstruct the Calendar. The latest and more accepted version was proposed by professor Rafael Tena (INAH),[444] based on the studies of Sahagún and Alfonso Caso (UNAM). His correlation confirms that the first day of the mexica year was February 13 of the old Julian calendar or February 23 of the current Gregorian calendar. Using the same count, it has been verified the date of the birth of Huitzilopochtli, the end of the year and a cycle or "Tie of the Years," and the New Fire Ceremony, day-sign "1 Tecpatl" of the year "2 Acatl,"[445] corresponding to the date February 22nd.

References

Aguilar-Moreno, Manuel (n.d.). "Aztec Art"[446] (PDF). *Aztec Art and Architecture*. Foundation for the Advancement of Mesoamerican Studies, Inc. (FAMSI). Archived from the original[447] (PDF) on 2008-06-25. Retrieved 2008-05-14.

Andrews, J. Richard (2003). *Introduction to Classical Nahuatl* (revised ed.). Norman: University of Oklahoma Press. ISBN 0-8061-3452-6. OCLC 50090230[448].

Aveni, Anthony F. (2000). *Empires of Time: Calendars, Clocks, and Cultures* (reprint of 1990 original ed.). London: Tauris Parke. ISBN 1-86064-602-6. OCLC 45144264[449].

Boone, Elizabeth Hill (1998). "Pictorial Documents and Visual Thinking in Postconquest Mexico"[450] (PDF Reprint). In Elizabeth Hill Boone; Tom Cubbins. *Native Traditions in the Postconquest World, A Symposium at Dumbarton Oaks 2nd through 4th October 1992*. Washington D.C.: Dumbarton Oaks Research Library and Collection. pp. 149–199. ISBN 0-88402-239-0. OCLC 34354931[451].

Boone, Elizabeth Hill (2000). *Stories in Red and Black: Pictorial Histories of the Aztec and Mixtec*. Austin: University of Texas Press. ISBN 0-292-70876-9. OCLC 40939882[452].

Boone, Elizabeth Hill (2007). *Cycles of Time and Meaning in the Mexican Books of Fate*. Joe R. and Teresa Lozano Long series in Latin American and Latino art and culture. Austin: University of Texas Press. ISBN 978-0-292-71263-8. OCLC 71632174[453].

Clavigero, Francesco Saverio (1807) [1787]. *The history of Mexico. Collected from Spanish and Mexican historians, from manuscripts, and ancient paintings of the Indians. Illustrated by charts, and other copper plates. To which are added, critical dissertations on the land, the animals, and inhabitants of Mexico, 2 vols.* Translated from the original Italian, by Charles Cullen, Esq. (2nd ed.). London: J. Johnson. OCLC 54014738[454].

Coe, Michael D. (1994) [1962]. *Mexico: from the Olmecs to the Aztecs* (4th, Revised and Enlarged ed.). London and New York: Thames & Hudson. ISBN 0-500-27722-2. OCLC 29708907[455].

Hassig, Ross (2001). *Time, History, and Belief in Aztec and Colonial Mexico*. Austin: University of Texas Press. ISBN 0-292-73139-6. OCLC 44167649[456].

Hernández de León-Portilla, Ascención (2004). "Lenguas y escrituras mesoamericanas"[457]. *Arqueología mexicana* (in Spanish). México, D.F.: Instituto Nacional de Antropología e Historia, Editorial Raíces. **12** (70): 20–25. ISSN 0188-8218[458]. Archived from the original[459] on 15 May 2008. Retrieved 2008-05-14.

Klein, Cecelia F. (2002). "La iconografía y el arte mesoamericano"[460] (PDF). *Arqueología mexicana* (in Spanish). México, D.F.: Instituto Nacional de Antropología e Historia, Editorial Raíces. **10** (55): 28–35. ISSN 0188-8218[458]. Archived from the original[461] (PDF) on 2006-01-03.

León-Portilla, Miguel (1963). *Aztec Thought and Culture: A Study of the Ancient Náhuatl Mind*. Civilization of the American Indian series, no. 67. Jack Emory Davis (trans.). Norman: University of Oklahoma Press. OCLC 181727[462].

Malmström, Vincent H. (1973-09-17). "Origin of the Mesoamerican 260-Day Calendar"[463] (PDF Reprinted). *Science*. Lancaster, PA: American Association for the Advancement of Science. **181** (4103): 939–941. Bibcode: 1973Sci...181..939M[464]. PMID 17835843[465]. doi: 10.1126/science.181.4103.939[466]. Archived[467] (PDF) from the original on 28 May 2008. Retrieved 2008-05-14.

Miller, Mary; Karl Taube (1993). *The Gods and Symbols of Ancient Mexico and the Maya: An Illustrated Dictionary of Mesoamerican Religion.* London: Thames & Hudson. ISBN 0-500-05068-6. OCLC 27667317[468].

Prem, Hanns J. (2008). *Manual de la antigua cronología Mexicana.* Mexico: Centro de Investigaciones y Estudios Superiores en Antropología Social. ISBN 978-968-496-694-9.

Read, Kay Almere (1998). *Time and Sacrifice in the Aztec Cosmos.* Bloomington: Indiana University Press. ISBN 0-253-33400-4. OCLC 37909790[469].

Sahagún, Bernardino de (1950–82) [ca. 1540–85]. *Florentine Codex: General History of the Things of New Spain, 13 vols. in 12.* vols. I-XII. Charles E. Dibble and Arthur J.O. Anderson (eds., trans., notes and illus.) (translation of *Historia General de las Cosas de la Nueva España* ed.). Santa Fe, NM and Salt Lake City: School of American Research and the University of Utah Press. ISBN 0-87480-082-X. OCLC 276351[470].

Smith, Michael E. (2003). *The Aztecs* (2nd ed.). Malden, MA: Blackwell Publishing. ISBN 0-631-23015-7. OCLC 48579073[471].

Townsend, Richard F. (2000). *The Aztecs* (Revised 2nd ed.). London: Thames & Hudson. ISBN 0-500-28132-7. OCLC 43337963[472].

Wimmer, Alexis (2006). "Dictionnaire de la langue nahuatl classique"[473] (online version, incorporating reproductions from *Dictionnaire de la langue nahuatl ou mexicaine* [1885], by Rémi Siméon). (in French) (in Nahuatl)

Zantwijk, Rudolph van (1985). *The Aztec Arrangement: The Social History of Pre-Spanish Mexico.* Norman: University of Oklahoma Press. ISBN 0-8061-1677-3. OCLC 11261299[474].

External links

- The Aztec Calendar - Ancient History Encyclopedia[475]
- (in Spanish) Detailed description of the temalacatl[476] from Mexico's Museo Nacional de Antropología
- Daily Aztec Calendar[477]

Astronomy in the medieval Islamic world

Islamic astronomy comprises the astronomical developments made in the Islamic world, particularly during the Islamic Golden Age (9th–13th centuries), and mostly written in the Arabic language. These developments mostly took place in the Middle East, Central Asia, Al-Andalus, and North Africa, and later in the Far East and India. It closely parallels the genesis of other Islamic sciences in its assimilation of foreign material and the amalgamation of the disparate elements of that material to create a science with Islamic characteristics. These included Greek, Sassanid, and Indian works in particular, which were translated and built upon.

Islamic astronomy played a significant role in the revival of Byzantine and European[478] astronomy following the loss of knowledge during the early medieval period, notably with the production of Latin translations of Arabic works during the 12th century. Islamic astronomy also had an influence on Chinese astronomy and Malian astronomy.

A significant number of stars in the sky, such as Aldebaran, Altair and Deneb, and astronomical terms such as alidade, azimuth, and nadir, are still referred to by their Arabic names. A large corpus of literature from Islamic astronomy remains today, numbering approximately 10,000 manuscripts scattered throughout the world, many of which have not been read or catalogued. Even so, a reasonably accurate picture of Islamic activity in the field of astronomy can be reconstructed.

History

Ahmad Dallal notes that, unlike the Babylonians, Greeks, and Indians, who had developed elaborate systems of mathematical astronomical study, the pre-Islamic Arabs relied entirely on empirical observations. These observations were based on the rising and setting of particular stars, and this area of astronomical study was known as *anwa*. *Anwa* continued to be developed after Islamization by the Arabs, where Islamic astronomers added mathematical methods to their empirical observations.[479] According to David King, after the rise of Islam, the religious obligation to determine the qibla and prayer times inspired more progress in astronomy for centuries.

Donald Hill (1993) divided Islamic Astronomy into the four following distinct time periods in its history:

Figure 49: *An 18th-century Persian astrolabe, kept at the Whipple Museum of the History of Science in Cambridge, England.*

Early Islam

Following the Islamic conquests, under the early caliphate, Muslim scholars began to absorb Hellenistic and Indian astronomical knowledge via translations into Arabic (in some cases via Persian).

The first astronomical texts that were translated into Arabic were of Indian and Persian origin. The most notable of the texts was *Zij al-Sindhind*,[480] an 8th-century Indian astronomical work that was translated by Muhammad ibn Ibrahim al-Fazari and Yaqub ibn Tariq after 770 CE with the assistance of Indian astronomers who visited the court of caliph Al-Mansur in 770. Another text translated was the *Zij al-Shah*, a collection of astronomical tables (based on Indian parameters) compiled in Sasanid Persia over two centuries. Fragments of texts during this period indicate that Arabs adopted the sine function (inherited from India) in place of the chords of arc used in Greek trigonometry.

Golden Age

The House of Wisdom was an academy established in Baghdad under Abbasid caliph Al-Ma'mun in the early 9th century. From this time, independent investigation into the Ptolemaic system became possible. According to Dallal

Figure 50: *A Large Persian Brass Celestial Globe with an ascription to Hadi Isfahani and a date of 1197 AH/ 1782-3 AD of typical spherical form, the globe engraved with markings, figures and astrological symbols, inscriptive details throughout*

(2010), the use of parameters, sources and calculation methods from different scientific traditions made the Ptolemaic tradition "receptive right from the beginning to the possibility of observational refinement and mathematical restructuring". Astronomical research was greatly supported by the Abbasid caliph al-Mamun through The House of Wisdom. Baghdad and Damascus became the centers of such activity. The caliphs not only supported this work financially, but endowed the work with formal prestige.Wikipedia:Citation needed

The first major Muslim work of astronomy was *Zij al-Sindh* by al-Khwarizmi in 830. The work contains tables for the movements of the sun, the moon and the five planets known at the time. The work is significant as it introduced Ptolemaic concepts into Islamic sciences. This work also marks the turning point in Islamic astronomy. Hitherto, Muslim astronomers had adopted a primarily research approach to the field, translating works of others and learning already discovered knowledge. Al-Khwarizmi's work marked the beginning of nontraditional methods of study and calculations.[481]

In 850, al-Farghani wrote *Kitab fi Jawani* (meaning "A compendium of the science of stars"). The book primarily gave a summary of Ptolemic cosmog-

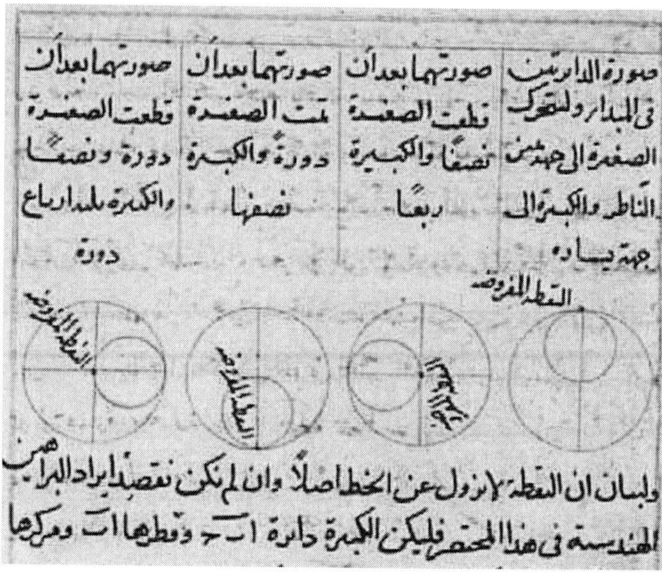

Figure 51: *The Tusi-couple is a mathematical device invented by Nasir al-Din al-Tusi in which a small circle rotates inside a larger circle twice the diameter of the smaller circle. Rotations of the circles cause a point on the circumference of the smaller circle to oscillate back and forth in linear motion along a diameter of the larger circle.*

raphy. However, it also corrected Ptolemy based on findings of earlier Arab astronomers. Al-Farghani gave revised values for the obliquity of the ecliptic, the precessional movement of the apogees of the sun and the moon, and the circumference of the earth. The book was widely circulated through the Muslim world, and even translated into Latin.[482]

The period when a distinctive Islamic system of astronomy flourished. The period began as the Muslim astronomers began questioning the framework of the Ptolemaic system of astronomy. These criticisms, however, remained within the geocentric framework and followed Ptolemy's astronomical paradigm; one historian described their work as "a reformist project intended to consolidate Ptolemaic astronomy by bringing it into line with its own principles."

Between 1025 and 1028, Ibn al-Haytham wrote his *Al-Shukuk ala Batlamyus* (meaning "Doubts on Ptolemy"). While maintaining the physical reality of the geocentric model, he criticized elements of the Ptolemic models. Many astronomers took up the challenge posed in this work, namely to develop alternate models that resolved these difficulties. In 1070, Abu Ubayd al-Juzjani

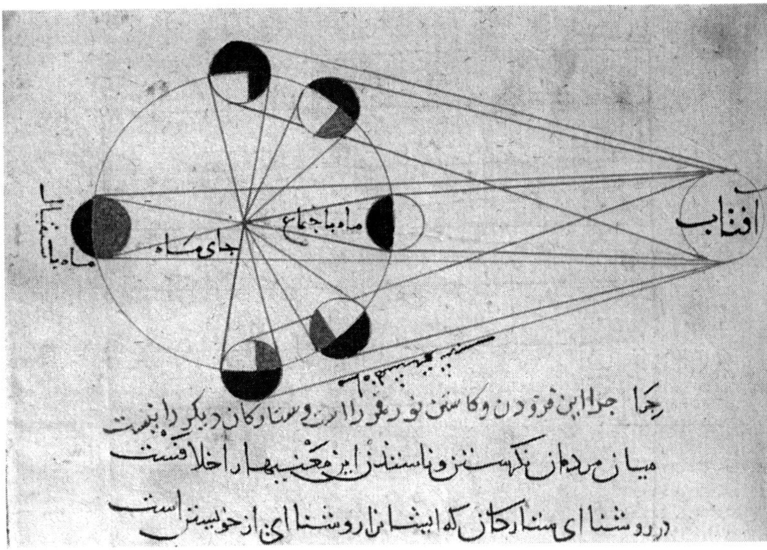

Figure 52: *An illustration from al-Biruni's astronomical works, explains the different phases of the moon.*

published the *Tarik al-Aflak*. In his work, he indicated the so-called "equant" problem of the Ptolemic model. Al-Juzjani even proposed a solution for the problem. In Al-Andalus, the anonymous work *al-Istidrak ala Batlamyus* (meaning "Recapitulation regarding Ptolemy"), included a list of objections to the Ptolemic astronomy.

Later period

Notable astronomers from the later medieval period include Mu'ayyad al-Din al-'Urdi (c. 1266), Nasir al-Din al-Tusi (1201–74), Qutb al-Din al Shirazi (c. 1311), Sadr al-Sharia al-Bukhari (c. 1347), Ibn al-Shatir (c. 1375), and Ali al-Qushji (c. 1474).[483]

In the fifteenth century, the Timurid ruler Ulugh Beg of Samarkand established his court as a center of patronage for astronomy. He himself studied it in his youth, and in 1420 ordered the construction of an observatory, which produced a new set of astronomical tables, as well as contributing to other scientific and mathematical advances.

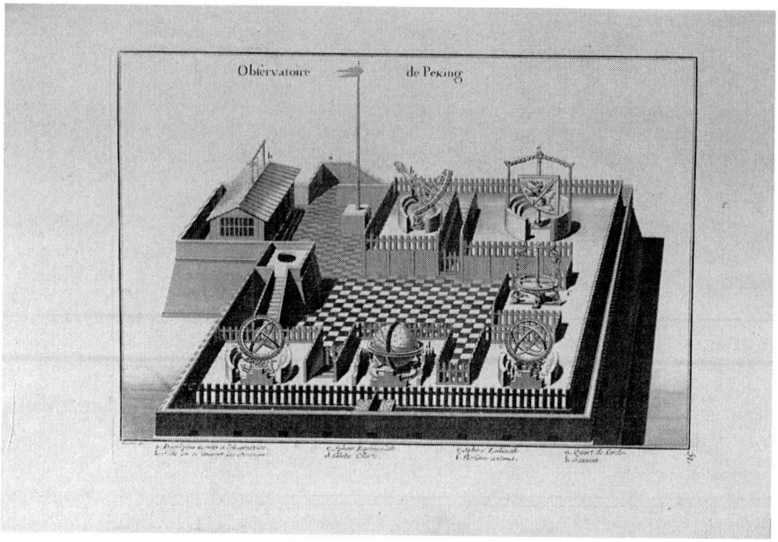

Figure 53: *Layout of the Beijing Ancient Observatory.*

Influences in East Asia

China

Islamic influence on Chinese astronomy was first recorded during the Song dynasty when a Hui Muslim astronomer named Ma Yize introduced the concept of 7 days in a week and made other contributions.

Islamic astronomers were brought to China in order to work on calendar making and astronomy during the Mongol Empire and the succeeding Yuan Dynasty. The Chinese scholar Yeh-lu Chu'tsai accompanied Genghis Khan to Persia in 1210 and studied their calendar for use in the Mongol Empire. Kublai Khan brought Iranians to Beijing to construct an observatory and an institution for astronomical studies.[484]

Several Chinese astronomers worked at the Maragheh observatory, founded by Nasir al-Din al-Tusi in 1259 under the patronage of Hulagu Khan in Persia. One of these Chinese astronomers was Fu Mengchi, or Fu Mezhai. In 1267, the Persian astronomer Jamal ad-Din, who previously worked at Maragha observatory, presented Kublai Khan with seven Persian astronomical instruments, including a terrestrial globe and an armillary sphere, as well as an astronomical almanac, which was later known in China as the *Wannian Li* ("Ten Thousand Year Calendar" or "Eternal Calendar"). He was known as "Zhamaluding" in China, where, in 1271, he was appointed by Khan as the first director of the Islamic observatory in Beijing, known as the Islamic Astronomical Bureau, which operated alongside the Chinese Astronomical Bureau for four centuries. Islamic astronomy gained a good reputation in China for its theory of planetary latitudes, which did not exist in Chinese astronomy at the time, and for its accurate prediction of eclipses.

Some of the astronomical instruments constructed by the famous Chinese astronomer Guo Shoujing shortly afterwards resemble the style of instrumentation built at Maragheh. In particular, the "simplified instrument" (*jianyi*) and the large gnomon at the Gaocheng Astronomical Observatory show traces of Islamic influence. While formulating the Shoushili calendar in 1281, Shoujing's work in spherical trigonometry may have also been partially influenced by Islamic mathematics, which was largely accepted at Kublai's court.[485] These possible influences include a pseudo-geometrical method for converting between equatorial and ecliptic coordinates, the systematic use of decimals in the underlying parameters, and the application of cubic interpolation in the calculation of the irregularity in the planetary motions.

Hongwu Emperor (r. 1368-1398) of the Ming Dynasty (1328–1398), in the first year of his reign (1368), conscripted Han and non-Han astrology specialists from the astronomical institutions in Beijing of the former Mongolian Yuan to Nanjing to become officials of the newly established national observatory.

That year, the Ming government summoned for the first time the astronomical officials to come south from the upper capital of Yuan. There were fourteen of them. In order to enhance accuracy in methods of observation and computation, Hongwu Emperor reinforced the adoption of parallel calendar systems, the Han and the Hui. In the following years, the Ming Court appointed several Hui astrologers to hold high positions in the Imperial Observatory. They wrote many books on Islamic astronomy and also manufactured astronomical equipment based on the Islamic system.

The translation of two important works into Chinese was completed in 1383: Zij (1366) and al-Madkhal fi Sina'at Ahkam al-Nujum, Introduction to Astrology (1004).

In 1384, a Chinese astrolabe was made for observing stars based on the instructions for making multi-purposed Islamic equipment. In 1385, the apparatus was installed on a hill in northern Nanjing.

Around 1384, during the Ming Dynasty, Hongwu Emperor ordered the Chinese translation and compilation of Islamic astronomical tables, a task that was carried out by the scholars Mashayihei, a Muslim astronomer, and Wu Bozong, a Chinese scholar-official. These tables came to be known as the *Huihui Lifa* (*Muslim System of Calendrical Astronomy*), which was published in China a

Figure 54: *Korean celestial globe based on the Huihui Lifa.*

number of times until the early 18th century, though the Qing Dynasty had officially abandoned the tradition of Chinese-Islamic astronomy in 1659. The Muslim astronomer Yang Guangxian was known for his attacks on the Jesuit's astronomical sciences.

Korea

In the early Joseon period, the Islamic calendar served as a basis for calendar reform owing to its superior accuracy over the existing Chinese-based calendars. A Korean translation of the *Huihui Lifa*, a text combining Chinese astronomy with Islamic astronomy works of Jamal ad-Din, was studied in Korea under the Joseon Dynasty during the time of Sejong in the 15th century. The tradition of Chinese-Islamic astronomy survived in Korea up until the early 19th century.

Observatories

The first systematic observations in Islam are reported to have taken place under the patronage of al-Mamun. Here, and in many other private observatories from Damascus to Baghdad, meridian degrees were measured, solar parameters were established, and detailed observations of the Sun, Moon, and planets were undertaken.

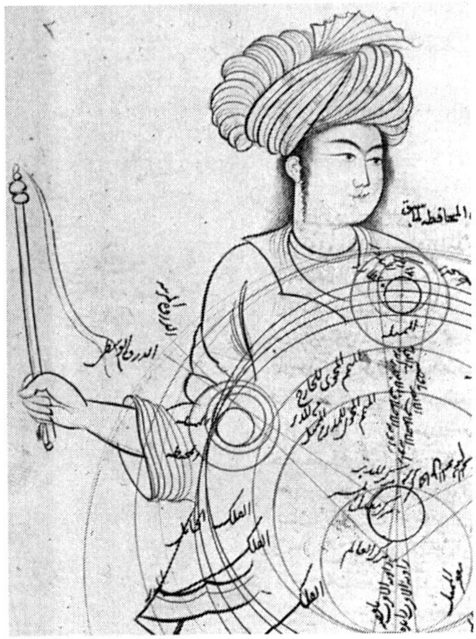

Figure 55: *Medieval manuscript by Qutb al-Din al-Shirazi depicting an epicyclic planetary model.*

In the 10th century, the Buwayhid dynasty encouraged the undertaking of extensive works in astronomy, such as the construction of a large-scale instrument with which observations were made in the year 950. We know of this by recordings made in the zij of astronomers such as Ibn al-Alam. The great astronomer Abd Al-Rahman Al Sufi was patronised by prince Adud o-dowleh, who systematically revised Ptolemy's catalogue of stars. Sharaf al-Daula also established a similar observatory in Baghdad. And reports by Ibn Yunus and al-Zarqall in Toledo and Cordoba indicate the use of sophisticated instruments for their time.

It was Malik Shah I who established the first large observatory, probably in Isfahan. It was here where Omar Khayyám with many other collaborators constructed a zij and formulated the Persian Solar Calendar a.k.a. the *jalali calendar*. A modern version of this calendar is still in official use in Iran today.

The most influential observatory was however founded by Hulegu Khan during the 13th century. Here, Nasir al-Din al-Tusi supervised its technical construction at Maragha. The facility contained resting quarters for Hulagu Khan, as well as a library and mosque. Some of the top astronomers of the day gathered there, and from their collaboration resulted important modifications to the Ptolemaic system over a period of 50 years.

Figure 56: *The Ulugh Beg Observatory in Samarqand.*

In 1420, prince Ulugh Beg, himself an astronomer and mathematician, founded another large observatory in Samarkand, the remains of which were excavated in 1908 by Russian teams.

And finally, Taqi al-Din Muhammad ibn Ma'ruf founded a large observatory in Ottoman Istanbul in 1577, which was on the same scale as those in Maragha and Samarkand. The observatory was short-lived however, as opponents of the observatory and prognostication from the heavens prevailed and the observatory was destroyed in 1580.[486] While the Ottoman clergy did not object to the science of astronomy, the observatory was primarily being used for astrology, which they did oppose, and successfully sought its destruction.

Instruments

Our knowledge of the instruments used by Muslim astronomers primarily comes from two sources: first the remaining instruments in private and museum collections today, and second the treatises and manuscripts preserved from the Middle Ages. Muslim astronomers of the "Golden Period" made many improvements to instruments already in use before their time, such as adding new scales or details.

Figure 57: *Work in the observatorium of Taqi al-Din.*

Celestial globes and armillary spheres

Celestial globes were used primarily for solving problems in celestial astronomy. Today, 126 such instruments remain worldwide, the oldest from the 11th century. The altitude of the sun, or the Right Ascension and Declination of stars could be calculated with these by inputting the location of the observer on the meridian ring[487] of the globe.

An armillary sphere had similar applications. No early Islamic armillary spheres survive, but several treatises on "the instrument with the rings" were written. In this context there is also an Islamic development, the spherical astrolabe, of which only one complete instrument, from the 14th century, has survived.

Astrolabes

Brass astrolabes were a Hellenistic invention. The first Islamic astronomer reported as having built an astrolabe is Muhammad al-Fazari (late 8th century).[488] Astrolabes were popular in the Islamic world during the "Golden Age", chiefly as an aid to finding the qibla. The earliest known example[489] is dated to 927/8 (AH 315).

The instruments were used to read the rise of the time of rise of the Sun and fixed stars. al-Zarqali of Andalusia constructed one such instrument in which,

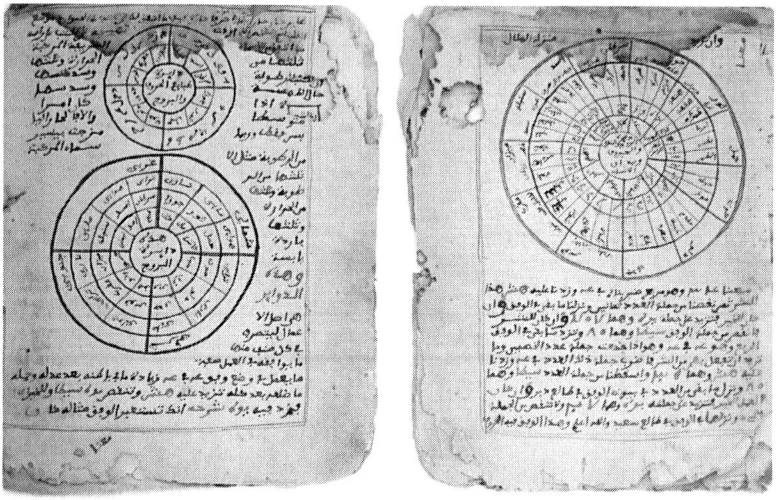

Figure 58: *The Timbuktu Manuscripts showing both mathematics and astronomy.*

unlike its predecessors, did not depend on the latitude of the observer, and could be used anywhere. This instrument became known in Europe as the Saphea[490].

Mechanical calendar

Abu Rayhan Biruni made an instrument he called "Box of the Moon", which was a mechanical lunisolar calendar, employing a gear train and eight gear-wheels. This was an early example of a fixed-wired knowledge processing machine.[491]

Sundials

Muslims made several important improvementsWikipedia:Avoid weasel words to the theory and construction of sundials, which they inherited from their Indian and Greek predecessors. Khwarizmi made tables for these instruments which considerably shortened the time needed to make specific calculations.

Sundials were frequently placed on mosques to determine the time of prayer. One of the most striking examples was built in the 14th century by the *muwaqqit* (timekeeper) of the Umayyad Mosque in Damascus, ibn al-Shatir.[492]

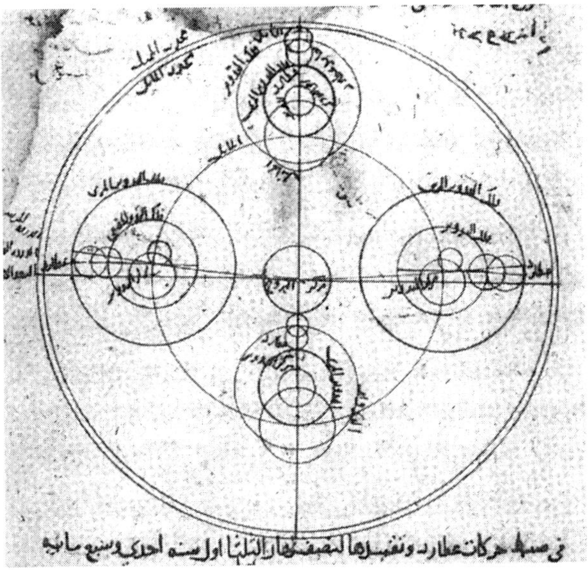

Figure 59: *Ibn al-Shatir's model for the appearances of Mercury, showing the multiplication of epicycles using the Tusi-couple, thus eliminating the Ptolemaic eccentrics and equant.*

Quadrants

Several forms of quadrants were invented by Muslims. Among them was the sine quadrant used for astronomical calculations and various forms of the horary quadrant, used to determine time (especially the times of prayer) by observations of the Sun or stars. A center of the development of quadrants was ninth-century Baghdad.[493]

Equatoriums

The Equatorium is an Islamic invention from Al-Andalus. The earliest known was probably made around 1015Wikipedia:Citation needed. It is a mechanical device for finding the positions of the Moon, Sun, and planets, without calculation using a geometrical model to represent the celestial body's mean and anomalistic position.

Notes and references

Bibliography

- Adnan, Abdulhak (1939), *La science chez les Turcs ottomans*, Paris
- Ajram, K. (1992), "Appendix B", *Miracle of Islamic Science*, Knowledge House Publishers, ISBN 0-911119-43-4
- Baker, A.; Chapter, L. (2002), "Part 4: The Sciences", *Philosophia Islamica*, in Sharif, M. M., "A History of Muslim Philosophy", *Philosophia Islamica*
- Covington, Richard, "Rediscovering Arabic science", *Saudi Aramco World* (May–June 2007 ed.), pp. 2–16Wikipedia:Identifying reliable sources
- Dallal, Ahmad (1999), "Science, Medicine and Technology", in Esposito, John, *The Oxford History of Islam*, Oxford University Press, New York
- Duhem, Pierre (1969) [First published 1908], *To Save the Phenomena: An Essay on the Idea of Physical theory from Plato to Galileo*, University of Chicago Press, Chicago, ISBN 0-226-16921-9
- Gautier, Antoine (December 2005), "L'âge d'or de l'astronomie ottomane", *L'Astronomie (*monthly magazine created by Camille Flammarion in 1882*)*, **119**
- Gill, M. (2005), *Was Muslim Astronomy the Harbinger of Copernicanism?*[494], archived from the original[495] on 2 January 2008, retrieved 2008-01-22
- Gingerich, Owen (April 1986), "Islamic astronomy"[496], *Scientific American*, **254** (10): 74, Bibcode: 1986SciAm.254...74G[497], doi: 10.1038/scientificamerican0486-74[498], archived from the original[499] on 2011-01-01, retrieved 2008-05-18
- Glick, Thomas F.; Livesey, Steven John; Wallis, Faith (2005), *Medieval Science, Technology, and Medicine: An Encyclopedia*, Routledge, ISBN 0-415-96930-1
- Hassan, Ahmad Y., *Transfer Of Islamic Technology To The West, Part II: Transmission Of Islamic Engineering*[500], archived from the original[501] on 18 February 2008, retrieved 2008-01-22
- Hill, Donald R. (1985), "Al-Biruni's mechanical calendar", *Annals of Science*, **42** (2): 139–163, doi: 10.1080/00033798500200141[502]
- Hill, Donald R. (May 1991), "Mechanical Engineering in the Medieval Near East", *Scientific American*: 64–69 (cf. Hill, Donald R., *Mechanical Engineering*[503], ISBN 0-07-037863-0, archived[504] from the original on 25 December 2007, retrieved 2008-01-22)
- Hill, Donald R. (1993), *Islamic Science And Engineering*, Edinburgh University Press, ISBN 0-7486-0455-3

- Huff, Toby (2003), *The Rise of Early Modern Science: Islam, China, and the West*, Cambridge University Press, ISBN 0-521-52994-8
- Ilyas, Mohammad (1996), *Islamic Astronomy and Science Development: Glorious Past, Challenging Future*, Pelanduk Publications, ISBN 967-978-549-1
- Iqbal, Muzaffar (2003), "Review: *World-Maps for Finding the Direction and Distance to Mecca: Innovation and Tradition in Islamic Science* by David A. King", *Islam & Science*, June 2003
- Iqbal, Muzaffar; Berjak, Rafik (2003), "Ibn Sina–Al-Biruni correspondence", *Islam & Science*, June 2003
- Kennedy, Edward S. (1947), "Al-Kashi's Plate of Conjunctions", *Isis*, **38** (1–2): 56–59, doi: 10.1086/348036[505]
- Kennedy, Edward S. (1950), "A Fifteenth-Century Planetary Computer: al-Kashi's "Tabaq al-Manateq" I. Motion of the Sun and Moon in Longitude", *Isis*, **41** (2): 180–183, doi: 10.1086/349146[506]
- Kennedy, Edward S. (1951), "An Islamic Computer for Planetary Latitudes", *Journal of the American Oriental Society*, Journal of the American Oriental Society, Vol. 71, No. 1, **71** (1): 13–21, JSTOR 595221[507], doi: 10.2307/595221[508]
- Kennedy, Edward S. (1952), "A Fifteenth-Century Planetary Computer: al-Kashi's "Tabaq al-Maneteq" II: Longitudes, Distances, and Equations of the Planets", *Isis*, **43** (1): 42–50, doi: 10.1086/349363[509]
- Kennedy, Edward S. (1956), "A Survey of Islamic Astronomical Tables", *Transactions of the American Philosophical Society*, Transactions of the American Philosophical Society, Vol. 46, No. 2, **46** (2): 123, JSTOR 1005726[510], doi: 10.2307/1005726[511]
- Kennedy, Edward S. (1961), "Al-Kashi's Treatise on Astronomical Observational Instruments", *Journal of Near Eastern Studies*, **20** (2): 98–108, doi: 10.1086/371617[512]
- Kennedy, Edward S. (1962), "Review: *The Observatory in Islam and Its Place in the General History of the Observatory* by Aydin Sayili", *Isis*, **53** (2): 237–239, doi: 10.1086/349558[513]
- Kennedy, Edward S. (1998), *Astronomy and Astrology in the Medieval Islamic World*, Brookfield, VT: Ashgate, ISBN 0-86078-682-X
- King, David A. (1983), "The Astronomy of the Mamluks", *Isis*, **74** (4): 531–555, doi: 10.1086/353360[514]
- King, David A. (1986), *Islamic mathematical astronomy*, London, ISBN 0-86078-407-X
- King, David A. (1997), "Two Iranian World Maps for Finding the Direction and Distance to Mecca", *Imago Mundi*, **49** (1): 62–82, doi: 10.1080/03085699708592859[515]
- King, David A. (1999a), "Islamic Astronomy", in Walker, Christopher,

Astronomy before the telescope, British Museum Press, pp. 143–174, ISBN 0-7141-2733-7

- King, David A. (1999b), *World-maps for Finding the Direction and Distance to Mecca: Innovation and Tradition in Islamic Science*, Brill Publishers, ISBN 90-04-11367-3
- King, David A. (2002), "A Vetustissimus Arabic Text on the Quadrans Vetus", *Journal for the History of Astronomy*, **33**: 237–255, Bibcode: 2002JHA....33..237K[516]
- King, David A. (December 2003), "14th-Century England or 9th-Century Baghdad? New Insights on the Elusive Astronomical Instrument Called Navicula de Venetiis", *Centaurus*, **45** (1–4): 204–226, doi: 10.1111/j.1600-0498.2003.450117.x[517]
- King, David A. (2004), "Reflections on some new studies on applied science in Islamic societies (8th–19th centuries)", *Islam & Science*, June 2004
- King, David A. (2005), *In Synchrony with the Heavens, Studies in Astronomical Timekeeping and Instrumentation in Medieval Islamic Civilization: Instruments of Mass Calculation*, Brill Publishers, ISBN 90-04-14188-X
- King, David A.; Cleempoel, Koenraad Van; Moreno, Roberto (2002), "A Recently Discovered Sixteenth-Century Spanish Astrolabe", *Annals of Science*, **59** (4): 331–362, doi: 10.1080/00033790110095813[518]
- Langermann, Y. Tzvi, ed. and trans. (1990), *Ibn al-Haytham's On the Configuration of the World*, Harvard Dissertations in the History of Science, New York: Garland, ISBN 0-8240-0041-2
- Marmura, Michael E.; Nasr, Seyyed Hossein (1965), "Review: *An Introduction to Islamic Cosmological Doctrines. Conceptions of Nature and Methods Used for Its Study by the Ikhwan Al-Safa'an, Al-Biruni, and Ibn Sina* by Seyyed Hossein Nasr", *Speculum*, Speculum, Vol. 40, No. 4, **40** (4): 744–746, JSTOR 2851429[519], doi: 10.2307/2851429[520]
- Marshall, O. S. (1950), "Alhazen and the Telescope", *Astronomical Society of the Pacific Leaflets*, **6**: 4
- Morrison, James E. (2007), *The Astrolabe*, Janus, ISBN 978-0-939320-30-1
- Nasr, Seyyed H. (1993), *An Introduction to Islamic Cosmological Doctrines* (2nd ed.), 1st edition by Harvard University Press, 2nd edition by State University of New York Press, ISBN 0-7914-1515-5
- Ragep, F. Jamil (2001a), "Tusi and Copernicus: The Earth's Motion in Context", *Science in Context*, Cambridge University Press, **14** (1–2): 145–163, doi: 10.1017/s0269889701000060[521]
- Ragep, F. Jamil (2001b), "Freeing Astronomy from Philosophy: An Aspect of Islamic Influence on Science", *Osiris, 2nd Series*, **16** (Science

in Theistic Contexts: Cognitive Dimensions): 49–64 & 66–71, Bibcode: 2001Osir...16...49R[522], doi: 10.1086/649338[523]

- Ragep, F. Jamil; Teresi, Dick; Hart, Roger (2002), *Ancient Roots of Modern Science*[524], *Talk of the Nation* (National Public Radio discussion; astronomy is discussed in the first fifteen-minute segment), archived[525] from the original on 1 January 2008, retrieved 2008-01-22
- Rashed, Roshdi; Morelon, Régis (1996), *Encyclopedia of the History of Arabic Science*, 1, **& 3**, Routledge, ISBN 0-415-12410-7
- Rashed, Roshdi (2007), "The Celestial Kinematics of Ibn al-Haytham", *Arabic Sciences and Philosophy*, Cambridge University Press, **17** (01): 7–55, doi: 10.1017/S0957423907000355[526]
- Rosen, Edward (1985), "The Dissolution of the Solid Celestial Spheres", *Journal of the History of Ideas*, Journal of the History of Ideas, Vol. 46, No. 1, **46** (1): 13–31, JSTOR 2709773[527], doi: 10.2307/2709773[528]
- Sabra, A. I. (1998), "Configuring the Universe: Aporetic, Problem Solving, and Kinematic Modeling as Themes of Arabic Astronomy", *Perspectives on Science*, **6**: 288–330
- Saliba, George (1979), "The First Non-Ptolemaic Astronomy at the Maraghah School", *Isis*, **70** (4): 571–576, doi: 10.1086/352344[529]
- Saliba, George (1980), "Al-Biruni", in Strayer, Joseph, *Dictionary of the Middle Ages*, **2**, Charles Scribner's Sons, New York
- Saliba, George; Sezgin, F. (1981), "Review: *Geschichte des arabischen Schriftiums. Band VI: Astronomie bis ca. 430 H* by F. Sezgin", *Journal of the American Oriental Society*, Journal of the American Oriental Society, Vol. 101, No. 2, **101** (2): 219–221, JSTOR 601763[530], doi: 10.2307/601763[531]
- Saliba, George (1994a), "Early Arabic Critique of Ptolemaic Cosmology: A Ninth-Century Text on the Motion of the Celestial Spheres", *Journal for the History of Astronomy*, **25**: 115–141, Bibcode: 1994JHA....25..115S[532]
- Saliba, George (1994b), *A History of Arabic Astronomy: Planetary Theories During the Golden Age of Islam*, New York University Press, ISBN 0-8147-8023-7
- Saliba, George (1999), *Whose Science is Arabic Science in Renaissance Europe?*[533], Columbia University, archived[534] from the original on 15 January 2008, retrieved 2008-01-22
- Saliba, George (2000), "Arabic versus Greek Astronomy: A Debate over the Foundations of Science", *Perspectives on Science*, **8** (4): 328–341, doi: 10.1162/106361400753373713[535]
- Saliba, George (2007), *Lecture at SOAS, London – Part 3/7*[536], Muslim Heritage & YouTube, retrieved 2008-01-22

- Singer, C. (1959), *A Short History of Scientific Ideas*, Oxford University Press, ISBN 0-19-881049-0
- Suter, H. (1902), *Mathematiker und Astronomen der Araber*
- Tabatabaei, Seyyed Muhammad Husayn, *Tafsir al-Mizan*
- Wickens, G. M. (1976), "The Middle East as a world Centre of science and medicine", in Savory, Roger M., *Introduction to Islamic Civilization*, Cambridge University Press, pp. 111–118, ISBN 0-521-09948-X

External links

- "Tubitak Turkish National Observatory Antalya"[537]
- "Scientific American" article on Islamic Astronomy[538]
- The Arab Union for Astronomy and Space Sciences (AUASS)[539]
- King Abdul Aziz Observatory[540]
- History of Islamic Astrolabes[541]
- Al-Sufi's constellations[542]

Renaissance Period

Astronomia nova

Astronomia nova (English: New Astronomy)

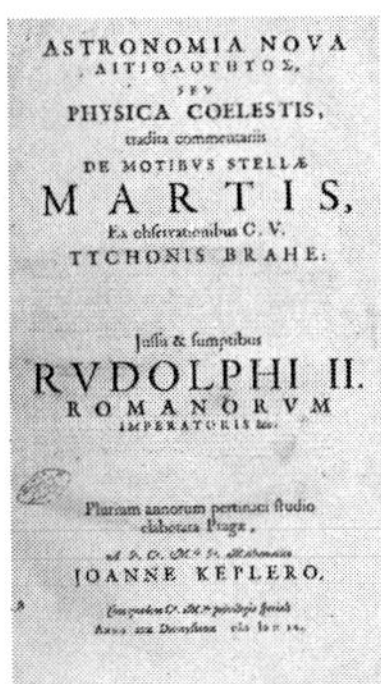
ASTRONOMIA NOVA
ΑΙΤΙΟΛΟΓΗΤΟΣ,
SEV
PHYSICA COELESTIS,
tradita commentariis
DE MOTIBVS STELLÆ
MARTIS,
Ex observationibus G. V.
TYCHONIS BRAHE:
Jussu & sumptibus
RVDOLPHI II.
ROMANORVM
IMPERATORIS &c:
Plurium annorum pertinaci studio elaborata Pragæ,
JOANNE KEPLERO,

Author	Johannes Kepler
Original title	*Astronomia Nova ΑΙΤΙΟΛΟΓΗΤΟΣ seu physica coelestis, tradita commentariis de motibus stellae Martis ex observationibus G.V. Tychonis Brahe (English: New Astronomy, Based upon Causes, or Celestial Physics, Treated by Means of Commentaries on the Motions of the Star Mars, from the Observations of Tycho Brahe, Gent)*
Language	Latin
Subject	astronomy

Astronomia nova (English: *New Astronomy*, full title in original Latin: *Astronomia Nova ΑΙΤΙΟΛΟΓΗΤΟΣ seu physica coelestis, tradita commentariis de motibus stellae Martis ex observationibus G.V. Tychonis Brahe*)[543,544] is a book, published in 1609, that contains the results of the astronomer Johannes Kepler's ten-year-long investigation of the motion of Mars. One of the most significant books in the history of astronomy, the *Astronomia nova* provided strong arguments for heliocentrism and contributed valuable insight into the movement of the planets. This included the first mention of the planets' elliptical paths and the change of their movement to the movement of free floating bodies as opposed to objects on rotating spheres. It is recognized as one of the most important works of the scientific revolution.

Background

Prior to Kepler, Nicolaus Copernicus proposed in 1543 that the Earth and other planets orbit the Sun. The Copernican model of the solar system was regarded Wikipedia:Citation needed as a device to explain the observed positions of the planets rather than a physical description.

Kepler sought for and proposed physical causes for planetary motion. His work is primarily based on the research of his mentor, Tycho Brahe. The two, though close in their work, had a tumultuous relationship. Regardless, on his deathbed, Brahe asked Kepler to make sure that he did not "die in vain," and to continue the development of his Tychonic system. Kepler would instead write the *Astronomia nova*, in which he rejects the Tychonic system, as well as the Ptolemaic system and the Copernican system. Some scholars have speculated that Kepler's dislike for Brahe may have had a hand in his rejection of the Tychonic system and formation of a new one.

Structure and Summary of the *Astronomia nova*

In English, the full title of his work is the ***New Astronomy, Based upon Causes, or Celestial Physics, Treated by Means of Commentaries on the Motions of the Star Mars, from the Observations of Tycho Brahe, Gent***. For over 650 pages, Kepler walks his readers, step by step, through his process of discovery so as to dispel any impression of "cultivating novelty," he says.

The *Astronomia nova*'s introduction, specifically the discussion of scripture, was the most widely distributed of Kepler's works in the seventeenth century. The introduction outlines the four steps Kepler took during his research. The first is his claim that the sun itself and not any imaginary point near the sun (as in the Copernican system) is the point where all the planes of the eccentrics of the planets intersect, or the center of the orbits of the planets. The second step consists of Kepler placing the sun as the center and mover of the other planets. This step also contains Kepler's reply to objections against placing the sun at the center of the universe, including objections based on scripture. In reply to scripture, he argues that it is not meant to claim physical dogma, and the content should be taken spiritually. In the third step, he posits that the sun

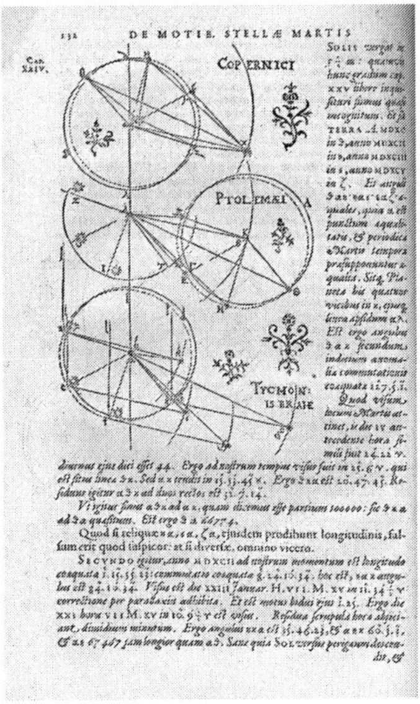

Figure 60: *Diagrams of the three models of planetary motion prior to Kepler*

is the source of the motion of all planets, using Brahe's proof based on comets that planets do not rotate on orbs. The fourth step consists of describing the path of planets as not a circle, but an oval.

As the *Astronomia nova* proper starts, Kepler demonstrates that the Tychonic, Ptolemaic, and Copernican systems are indistinguishable on the basis of observations alone. The three models predict the same positions for the planets in the near term, although they diverge from historical observations, and fail in their ability to predict future planetary positions by a small, though absolutely measurable amount. Kepler here introduces his famous diagram of the movement of Mars in relation to Earth if Earth remained unmoving at the center of its orbit. The diagram shows that Mars's orbit would be completely imperfect and never follow along the same path.

Kepler discusses all his work at great length throughout the book. He addresses this length in the sixteenth chapter:

> *If thou art bored with this wearisome method of calculation, take pity on me, who had to go through with at least seventy repetitions of it, at a very great loss of time.*

Kepler, in a very important step, also questions the assumption that the planets move around the center of their orbit at a uniform rate. He finds that computing critical measurements based upon the Sun's actual position in the sky, instead of the Sun's "mean" position injects a significant degree of uncertainty into the models, opening the path for further investigations. The idea that the planets do not move at a uniform rate, but at a speed that varies as their distance from the Sun, was completely revolutionary and would become his second law (discovered before his first). Kepler, in his calculations leading to his second law, made multiple mathematical errors, which luckily cancelled each other out "as if by miracle."

Given this second law, he puts forth in Chapter 33 that the sun is the engine that moves the planets. To describe the motion of the planets, he claims the sun emits a physical species, analogous to the light it also emits, which pushes the planets along. He also suggests a second force within every planet itself that pulls it towards the sun to keep it from spiraling off into space.

Kepler then attempts to finally find the true path of the planets, which he determines is an ellipse. His initial attempt to define the orbit of Mars, far before he arrived at the ellipse shape, was off by only eight minutes of arc, yet this was enough for Kepler to require an entirely new system. Kepler tried a number of shapes before the ellipse, including an egg shape. What is more, he discovered the mathematical definition for the ellipse as the orbit, then rejected it, then adopted the ellipse without knowing that it was the same:

> *"I laid [the original equation] aside, and fell back on ellipses, believing that this was quite a different hypothesis, whereas the two, as I shall prove in the next chapter, are one in the same...Ah, what a foolish bird I have been!"*

Kepler's laws

The *Astronomia nova* records the discovery of the first two of the three principles known today as Kepler's laws of planetary motion, which are:

1. That the planets move in elliptical orbits with the sun at one focus.[545]
2. That the speed of the planet changes at each moment such that the time between two positions is always proportional to the area swept out on the orbit between these positions.[546]

Kepler discovered the "second law" before the first. He presented his second law in two different forms: In Chapter 32 he states that the speed of the planet varies inversely based upon its distance from the Sun, and therefore he could measure changes in position of the planet by adding up all the distance measures, or looking at the area along an orbital arc. This is his so-called "distance law". In Chapter 59, he states that a radius from the Sun to a planet sweeps out equal areas in equal times. This is his so-called "area law".

However, Kepler's "area-time principle" did not facilitate easy calculation of planetary positions. Kepler could divide up the orbit into an arbitrary number of parts, compute the planet's position for each one of these, and then refer all questions to a table, but he could not determine the position of the planet at each and every individual moment because the speed of the planet was always changing. This paradox, referred to as the "Kepler problem," prompted the development of calculus.

The "third law"

Kepler discovered his "third law" a decade after the publication of the *Astronomia nova* as a result of his investigations in the 1619 *Harmonices Mundi* (*Harmonies of the world*).[547] He found that the ratio of the cube of the length of the semi-major axis of each planet's orbit, to the square of time of its orbital period, is the same for all planets.

Kepler's knowledge of gravity

In his introductory discussion of a moving earth, Kepler addressed the question of how the Earth could hold its parts together if it moved away from the center of the universe which, according to Aristotelian physics, was the place toward which all heavy bodies naturally moved. Kepler proposed an attractive force similar to magnetism, which may have been known by Newton.

> *"Gravity is a mutual corporeal disposition among kindred bodies to unite or join together; thus the earth attracts a stone much more than the stone seeks the earth. (The magnetic faculty is another example of this sort).... If two stones were set near one another in some place in the world outside the sphere of influence of a third kindred body, these stones, like two magnetic bodies, would come together in an intermediate place, each approaching the other by a space proportional to the bulk [moles] of the other.... For it follows that if the earth's power of attraction will be much more likely to extend to the moon and far beyond, and accordingly, that nothing that consists to any extent whatever of terrestrial material, carried up on high, ever escapes the grasp of this mighty power of attraction."*

Kepler discusses the moon's gravitational effect upon the tides as follows:[548]

> *The sphere of the attractive virtue which is in the moon extends as far as the earth, and entices up the waters; but as the moon flies rapidly across the zenith, and the waters cannot follow so quickly, a flow of the ocean is occasioned in the torrid zone towards the westward. If the attractive virtue of the moon extends as far as the earth, it follows with greater reason that the attractive virtue of the earth extends as far as the moon and much*

farther; and, in short, nothing which consists of earthly substance anyhow constituted although thrown up to any height, can ever escape the powerful operation of this attractive virtue.

Kepler also clarifies the concept of lightness in terms of relative density, in opposition to the Aristotelian concept of the absolute nature or quality of lightness as follows. His argument could easily be applied today to something like the flight of a hot air balloon.

Nothing which consists of corporeal matter is absolutely light, but that is comparatively lighter which is rarer, either by its own nature, or by accidental heat. And it is not to be thought that light bodies are escaping to the surface of the universe while they are carried upwards, or that they are not attracted by the earth. They are attracted, but in a less degree, and so are driven outwards by the heavy bodies; which being done, they stop, and are kept by the earth in their own place.

In reference to Kepler's discussion relating to gravitation, Walter William Bryant makes the following statement in his book *Kepler* (1920).

...the Introduction to Kepler's "Commentaries on the Motion of Mars," always regarded as his most valuable work, must have been known to Newton, so that no such incident as the fall of an apple was required to provide a necessary and sufficient explanation of the genesis of his Theory of Universal Gravitation. Kepler's glimpse at such a theory could have been no more than a glimpse, for he went no further with it. This seems a pity, as it is far less fanciful than many of his ideas, though not free from the "virtues" and "animal faculties," that correspond to Gilbert's "spirits and humours".

Kepler considered that this attraction was mutual and was proportional to the bulk of the bodies, but he considered it to have a limited range and he did not consider whether or how this force may have varied with distance. Furthermore, this attraction only acted between "kindred bodies"—bodies of a similar nature, a nature which he did not clearly define. Kepler's idea differed significantly from Newton's later concept of gravitation and it can be "better thought of as an episode in the struggle for heliocentrism than as a step toward Universal gravitation."

Commemoration

The 2009 International Year of Astronomy commemorates the 400th anniversary of the publication of this work.

References

- Johannes Kepler, *New Astronomy,* translated by William H. Donahue, Cambridge: Cambridge Univ. Pr., 1992. ISBN 0-521-30131-9
- https://web.archive.org/web/20080512144006/http://www.hao.ucar.edu/Public/education/bios/kepler.3.html
- https://web.archive.org/web/20060920211226/http://www.library.usyd.edu.au/libraries/rare/modernity/kepler4.html

External links

- *Astronomia nova*[549] by Johannes Kepler, 1609, in Latin, full text scan
- Origins of Modernity - Kepler: *Astronomia nova*[550]

Epitome Astronomiae Copernicanae

The ***Epitome Astronomiae Copernicanae*** was an influential astronomy book on the heliocentric system published by Johannes Kepler in the period 1617 to 1621. The first volume (books I–III) was printed in 1617, the second (book IV) in 1620, and the third (books V–VII) in 1621.

Content

The book contained in particular the first version in print of his third law of planetary motion. The work was intended as a textbook, and the first part was written by 1615. Divided into seven books, the *Epitome* covers much of Kepler's earlier thinking, as well as his later positions on physics, metaphysics and archetypes. In Book IV he supported the Copernican cosmology. Book V provided mathematics underpinning Kepler's views. Kepler wrote and published this work in parallel with his *Harmonices Mundi* (1619), the last Books V to VII appearing in 1621.

The term "inertia" was first introduced in the *Epitome*.[551]

The work was banned by the Catholic Church as heretical.

Editions

- 1635 reprint: *Epitome Astronomiae Copernicanae*[552], Vols. 1–3, Schönwetterus.

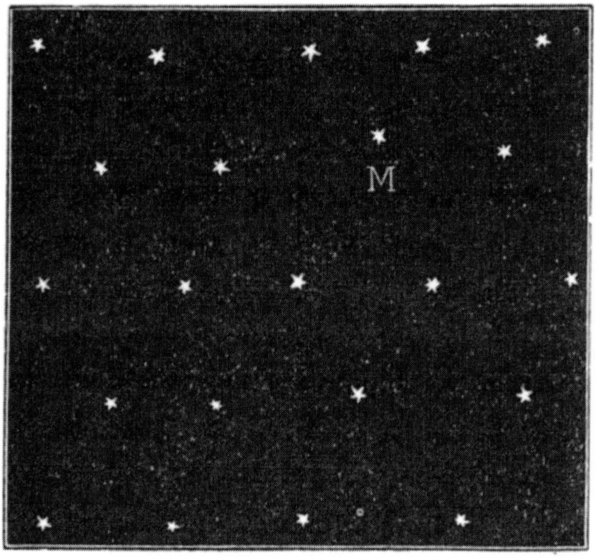

Figure 61: *Kepler's Figure 'M' from the Epitome, showing the world as belonging to just one of any number of similar stars*

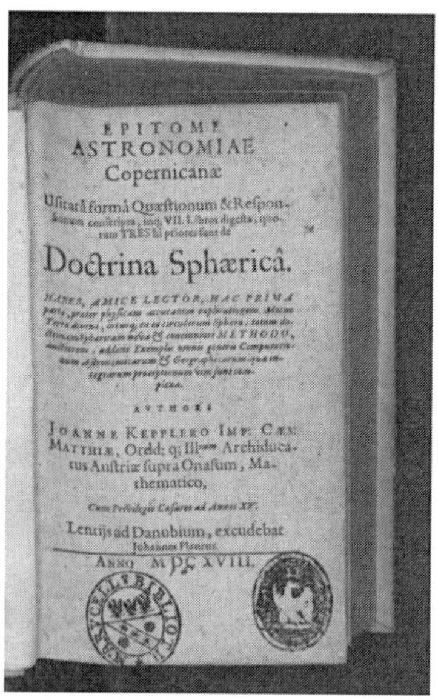

Figure 62: *Epitome astronomiae copernicanae (1618)*

Translations

- 1939: *Epitome of Copernican astronomy. Books IV and V, The organization of the world and the doctrine ...*; trans. by Charles Glenn Wallis. Annapolis: St John's Bookstore.
  - –ditto.–reissued with Ptolemy's Almagest. Chicago: Encyclopædia Britannica (1955).
- 1995: *Epitome of Copernican astronomy; & Harmonies of the world*; translated by Charles Glenn Wallis. Amherst: Prometheus Books.

Modern astronomy

Observational astronomy

Observational astronomy is a division of astronomy that is concerned with recording data about the observable universe, in contrast with theoretical astronomy, which is mainly concerned with calculating the measurable implications of physical models. It is the practice and study of observing celestial objects with the use of telescopes and other astronomical instruments.

As a science, the study of astronomy is somewhat hindered in that direct experiments with the properties of the distant universe are not possible. However, this is partly compensated by the fact that astronomers have a vast number of visible examples of stellar phenomena that can be examined. This allows for observational data to be plotted on graphs, and general trends recorded. Nearby examples of specific phenomena, such as variable stars, can then be used to infer the behavior of more distant representatives. Those distant yardsticks can then be employed to measure other phenomena in that neighborhood, including the distance to a galaxy.

Galileo Galilei turned a telescope to the heavens and recorded what he saw. Since that time, observational astronomy has made steady advances with each improvement in telescope technology.

Subdivisions of observational astronomy

A traditional division of observational astronomy is given by the region of the electromagnetic spectrum observed:

- Optical astronomy is the part of astronomy that uses optical components (mirrors, lenses and solid-state detectors) to observe light from near infrared to near ultraviolet wavelengths. Visible-light astronomy (using wavelengths that can be detected with the eyes, about 400 - 700 nm) falls in the middle of this range.

Figure 63: *Mayall telescope at Kitt Peak National Observatory*

Figure 64: *An assembly to observe meteors*

Figure 65: *Ultra HD photography taken at La Silla Observatory.*

- Infrared astronomy deals with the detection and analysis of infrared radiation (this typically refers to wavelengths longer than the detection limit of silicon solid-state detectors, about 1 μm wavelength). The most common tool is the reflecting telescope but with a detector sensitive to infrared wavelengths. Space telescopes are used at certain wavelengths where the atmosphere is opaque, or to eliminate noise (thermal radiation from the atmosphere).
- Radio astronomy detects radiation of millimetre to decametre wavelength. The receivers are similar to those used in radio broadcast transmission but much more sensitive. See also Radio telescopes.
- High-energy astronomy includes X-ray astronomy, gamma-ray astronomy, and extreme UV astronomy.

Methods

In addition to using electromagnetic radiation, modern astrophysicists can also make observations using neutrinos, cosmic rays or gravitational waves. Observing a source using multiple methods is known as multi-messenger astronomy.

Optical and radio astronomy can be performed with ground-based observatories, because the atmosphere is relatively transparent at the wavelengths being

Figure 66: *Sunset over Mauna Kea Observatories.*

detected. Observatories are usually located at high altitudes so as to minimise the absorption and distortion caused by the Earth's atmosphere. Some wavelengths of infrared light are heavily absorbed by water vapor, so many infrared observatories are located in dry places at high altitude, or in space.

The atmosphere is opaque at the wavelengths used by X-ray astronomy, gamma-ray astronomy, UV astronomy and (except for a few wavelength "windows") far infrared astronomy, so observations must be carried out mostly from balloons or space observatories. Powerful gamma rays can, however be detected by the large air showers they produce, and the study of cosmic rays is a rapidly expanding branch of astronomy.

Important factors

For much of the history of observational astronomy, almost all observation was performed in the visual spectrum with optical telescopes. While the Earth's atmosphere is relatively transparent in this portion of the electromagnetic spectrum, most telescope work is still dependent on seeing conditions and air transparency, and is generally restricted to the night time. The seeing conditions depend on the turbulence and thermal variations in the air. Locations that are frequently cloudy or suffer from atmospheric turbulence limit the resolution of observations. Likewise the presence of the full Moon can brighten up the sky with scattered light, hindering observation of faint objects.

For observation purposes, the optimal location for an optical telescope is undoubtedly in outer space. There the telescope can make observations without being affected by the atmosphere. However, at present it remains costly to lift telescopes into orbit. Thus the next best locations are certain mountain peaks that have a high number of cloudless days and generally possess good atmospheric conditions (with good seeing conditions). The peaks of the islands of Mauna Kea, Hawaii and La Palma possess these properties, as to a lesser extent do inland sites such as Llano de Chajnantor, Paranal, Cerro Tololo and La Silla in Chile. These observatory locations have attracted an assemblage of powerful telescopes, totalling many billion US dollars of investment.

The darkness of the night sky is an important factor in optical astronomy. With the size of cities and human populated areas ever expanding, the amount of artificial light at night has also increased. These artificial lights produce a diffuse background illumination that makes observation of faint astronomical features very difficult without special filters. In a few locations such as the state of Arizona and in the United Kingdom, this has led to campaigns for the reduction of light pollution. The use of hoods around street lights not only improves the amount of light directed toward the ground, but also helps reduce the light directed toward the sky.

Atmospheric effects (astronomical seeing) can severely hinder the resolution of a telescope. Without some means of correcting for the blurring effect of the shifting atmosphere, telescopes larger than about 15–20 cm in aperture can not achieve their theoretical resolution at visible wavelengths. As a result, the primary benefit of using very large telescopes has been the improved light-gathering capability, allowing very faint magnitudes to be observed. However the resolution handicap has begun to be overcome by adaptive optics, speckle imaging and interferometric imaging, as well as the use of space telescopes.

Measuring results

Astronomers have a number of observational tools that they can use to make measurements of the heavens. For objects that are relatively close to the Sun and Earth, direct and very precise position measurements can be made against a more distant (and thereby nearly stationary) background. Early observations of this nature were used to develop very precise orbital models of the various planets, and to determine their respective masses and gravitational perturbations. Such measurements led to the discovery of the planets Uranus, Neptune, and (indirectly) Pluto. They also resulted in an erroneous assumption of a fictional planet Vulcan within the orbit of Mercury (but the explanation of the precession of Mercury's orbit by Einstein is considered one of the triumphs of his general relativity theory).

Figure 67: *ALMA is the world's most powerful telescope for studying the Universe at submillimeter and millimeter wavelengths.*

Developments and diversity

In addition to examination of the universe in the optical spectrum, astronomers have increasingly been able to acquire information in other portions of the electromagnetic spectrum. The earliest such non-optical measurements were made of the thermal properties of the Sun. Instruments employed during a solar eclipse could be used to measure the radiation from the corona.

Radio astronomy

With the discovery of radio waves, radio astronomy began to emerge as a new discipline in astronomy. The long wavelengths of radio waves required much larger collecting dishes in order to make images with good resolution, and later led to the development of the multi-dish interferometer for making high-resolution aperture synthesis radio images (or "radio maps"). The development of the microwave horn receiver led to the discovery of the microwave background radiation associated with the big bang.

Radio astronomy has continued to expand its capabilities, even using radio astronomy satellites to produce interferometers with baselines much larger than the size of the Earth. However, the ever-expanding use of the radio spectrum for other uses is gradually drowning out the faint radio signals from the stars. For this reason, in the future radio astronomy might be performed from shielded locations, such as the far side of the Moon.

Figure 68: *Fully-steerable radio telescope in Green Bank, West Virginia.*

Late 20th century developments

The last part of the twentieth century saw rapid technological advances in astronomical instrumentation. Optical telescopes were growing ever larger, and employing adaptive optics to partly negate atmospheric blurring. New telescopes were launched into space, and began observing the universe in the infrared, ultraviolet, x-ray, and gamma ray parts of the electromagnetic spectrum, as well as observing cosmic rays. Interferometer arrays produced the first extremely high-resolution images using aperture synthesis at radio, infrared and optical wavelengths. Orbiting instruments such as the Hubble Space Telescope produced rapid advances in astronomical knowledge, acting as the workhorse for visible-light observations of faint objects. New space instruments under development are expected to directly observe planets around other stars, perhaps even some Earth-like worlds.

In addition to telescopes, astronomers have begun using other instruments to make observations.

Figure 69: *Skalnaté pleso observatory, Slovakia.*

Other instruments

Neutrino astronomy is the branch of astronomy that observes astronomical objects with neutrino detectors in special observatories, usually huge underground tanks. Nuclear reactions in stars and supernova explosions produce very large numbers of neutrinos, a very few of which may be detected by a neutrino telescope. Neutrino astronomy is motivated by the possibility of observing processes that are inaccessible to optical telescopes, such as the Sun's core.

Gravitational wave detectors are being designed that may capture events such as collisions of massive objects such as neutron stars or black holes.

Robotic spacecraft are also being increasingly used to make highly detailed observations of planets within the solar system, so that the field of planetary science now has significant cross-over with the disciplines of geology and meteorology.

Observation tools

Telescopes

The key instrument of nearly all modern observational astronomy is the telescope. This serves the dual purposes of gathering more light so that very faint

Figure 70: *One of the Oldest Observatories in South America is the Quito Astronomical Observatory, founded in 1873 and located 12 minutes south of the Equator in Quito, Ecuador. The Quito Astronomical Observatory is the National Observatory of Ecuador and is located in the Historic Center of Quito and is managed by the National Polytechnic School.*[553]

objects can be observed, and magnifying the image so that small and distant objects can be observed. Optical astronomy requires telescopes that use optical components of great precision. Typical requirements for grinding and polishing a curved mirror, for example, require the surface to be within a fraction of a wavelength of light of a particular conic shape. Many modern "telescopes" actually consist of arrays of telescopes working together to provide higher resolution through aperture synthesis.

Large telescopes are housed in domes, both to protect them from the weather and to stabilize the environmental conditions. For example, if the temperature is different from one side of the telescope to the other, the shape of the structure changes, due to thermal expansion pushing optical elements out of position. This can affect the image. For this reason, the domes are usually bright white (titanium dioxide) or unpainted metal. Domes are often opened around sunset, long before observing can begin, so that air can circulate and bring the entire telescope to the same temperature as the surroundings. To prevent wind-buffet or other vibrations affecting observations, it is standard practice to mount the

Figure 71: *An amateur astrophotography setup with an automated guide system connected to a laptop.*

telescope on a concrete pier whose foundations are entirely separate from those of the surrounding dome and building.

To do almost any scientific work requires that telescopes track objects as they wheel across the visible sky. In other words, they must smoothly compensate for the rotation of the Earth. Until the advent of computer controlled drive mechanisms, the standard solution was some form of equatorial mount, and for small telescopes this is still the norm. However, this is a structurally poor design and becomes more and more cumbersome as the diameter and weight of the telescope increases. The world's largest equatorial mounted telescope is the 200 inch (5.1 m) Hale Telescope, whereas recent 8–10 m telescopes use the structurally better altazimuth mount, and are actually physically *smaller* than the Hale, despite the larger mirrors. As of 2006, there are design projects underway for gigantic alt-az telescopes: the Thirty Metre Telescope[554], and the 100 m diameter Overwhelmingly Large Telescope.[555]

Amateur astronomers use such instruments as the Newtonian reflector, the Refractor and the increasingly popular Maksutov telescope.

Photography

The photograph has served a critical role in observational astronomy for over a century, but in the last 30 years it has been largely replaced for imaging applications by digital sensors such as CCDs and CMOS chips. Specialist areas of astronomy such as photometry and interferometry have utilised electronic detectors for a much longer period of time. Astrophotography uses specialised photographic film (or usually a glass plate coated with photographic emulsion), but there are a number of drawbacks, particularly a low quantum efficiency, of the order of 3%, whereas CCDs can be tuned for a QE >90% in a narrow band. Almost all modern telescope instruments are electronic arrays, and older telescopes have been either been retrofitted with these instruments or closed down. Glass plates are still used in some applications, such as surveying,Wikipedia:Citation needed because the resolution possible with a chemical film is much higher than any electronic detector yet constructed.

Advantages

Prior to the invention of photography, all astronomy was done with the naked eye. However, even before films became sensitive enough, scientific astronomy moved entirely to film, because of the overwhelming advantages:

- The human eye discards what it sees from split-second to split-second, but photographic film gathers more and more light for as long as the shutter is open.
- The resulting image is permanent, so many astronomers can use the same data.
- It is possible to see objects as they change over time (SN 1987A is a spectacular example).

Blink comparator

The blink comparator is an instrument that is used to compare two nearly identical photographs made of the same section of sky at different points in time. The comparator alternates illumination of the two plates, and any changes are revealed by blinking points or streaks. This instrument has been used to find asteroids, comets, and variable stars.

Micrometer

The position or cross-wire micrometer is an implement that has been used to measure double stars. This consists of a pair of fine, movable lines that can be moved together or apart. The telescope lens is lined up on the pair and oriented using position wires that lie at right angles to the star separation. The movable wires are then adjusted to match the two star positions. The separation of the stars is then read off the instrument, and their true separation determined based on the magnification of the instrument.

Figure 72: *50 cm refracting telescope at Nice Observatory.*

Spectrograph

A vital instrument of observational astronomy is the spectrograph. The absorption of specific wavelengths of light by elements allows specific properties of distant bodies to be observed. This capability has resulted in the discovery of the element of helium in the Sun's emission spectrum, and has allowed astronomers to determine a great deal of information concerning distant stars, galaxies, and other celestial bodies. Doppler shift (particularly "redshift") of spectra can also be used to determine the radial motion or distance with respect to the Earth.

Early spectrographs employed banks of prisms that split light into a broad spectrum. Later the grating spectrograph was developed, which reduced the amount of light loss compared to prisms and provided higher spectral resolution. The spectrum can be photographed in a long exposure, allowing the spectrum of faint objects (such as distant galaxies) to be measured.

Stellar photometry came into use in 1861 as a means of measuring stellar colors. This technique measured the magnitude of a star at specific frequency ranges, allowing a determination of the overall color, and therefore temperature of a star. By 1951 an internationally standardized system of UBV-magnitudes (*U*ltraviolet-*B*lue-*V*isual) was adopted.

Figure 73: *The main platform at La Silla hosts a huge range of telescopes with which astronomers can explore the Universe.*

Photoelectric photometry

Photoelectric photometry using the CCD is now frequently used to make observations through a telescope. These sensitive instruments can record the image nearly down to the level of individual photons, and can be designed to view in parts of the spectrum that are invisible to the eye. The ability to record the arrival of small numbers of photons over a period of time can allow a degree of computer correction for atmospheric effects, sharpening up the image. Multiple digital images can also be combined to further enhance the image. When combined with the adaptive optics technology, image quality can approach the theoretical resolution capability of the telescope.

Filters are used to view an object at particular frequencies or frequency ranges. Multilayer film filters can provide very precise control of the frequencies transmitted and blocked, so that, for example, objects can be viewed at a particular frequency emitted only by excited hydrogen atoms. Filters can also be used to partially compensate for the effects of light pollution by blocking out unwanted light. Polarization filters can also be used to determine if a source is emitting polarized light, and the orientation of the polarization.

Observing

Astronomers observe a wide range of astronomical sources, including high-redshift galaxies, AGNs, the afterglow from the Big Bang and many different types of stars and protostars.

A variety of data can be observed for each object. The position coordinates locate the object on the sky using the techniques of spherical astronomy, and the magnitude determines its brightness as seen from the Earth. The relative brightness in different parts of the spectrum yields information about the temperature and physics of the object. Photographs of the spectra allow the chemistry of the object to be examined.

Parallax shifts of a star against the background can be used to determine the distance, out to a limit imposed by the resolution of the instrument. The radial velocity of the star and changes in its position over time (proper motion) can be used to measure its velocity relative to the Sun. Variations in the brightness of the star give evidence of instabilities in the star's atmosphere, or else the presence of an occulting companion. The orbits of binary stars can be used to measure the relative masses of each companion, or the total mass of the system. Spectroscopic binaries can be found by observing doppler shifts in the spectrum of the star and its close companion.

Stars of identical masses that formed at the same time and under similar conditions typically have nearly identical observed properties. Observing a mass of closely associated stars, such as in a globular cluster, allows data to be assembled about the distribution of stellar types. These tables can then be used to infer the age of the association.

For distant galaxies and AGNs observations are made of the overall shape and properties of the galaxy, as well as the groupings where they are found. Observations of certain types of variable stars and supernovae of known luminosity, called standard candles, in other galaxies allows the inference of the distance to the host galaxy. The expansion of space causes the spectra of these galaxies to be shifted, depending on the distance, and modified by the Doppler effect of the galaxy's radial velocity. Both the size of the galaxy and its redshift can be used to infer something about the distance of the galaxy. Observations of large numbers of galaxies are referred to as redshift surveys, and are used to model the evolution of galaxy forms.

Related lists

- List of observatories
- List of radio telescopes

External links

	Wikiversity has learning resources about ***Observational astronomy***

Media related to Observational astronomy at Wikimedia Commons

Appendix

References

[1]Sinclair 2006:13
[2]Ruggles 2005:19
[3]Ruggles 1999:155
[4]Iwaniszewski 2003, 7-10
[5]Aveni 1980
[6]Chiu & Morrison 1980
[7]Magli 2008
[8]McCluskey 2005
[9]Carlson 1999
[10]Bostwick 2006:13
[11]Michell, 2001:9-10
[12]Johnson, 1912:225
[13]Hoskin, 2001:7
[14]Michell, 2001:17-18
[15]Sinclair 2006:17
[16]Ruggles 2005:312-3
[17]Sinclair 2006:8
[18]Mackie 2006:243
[19]Hawkins 1976
[20]Atkinson 1966
[21]Thom 1988:9-10
[22]MacKie 1977
[23]Gingerich 2000
[24]Krupp 1979:18
[25]Hicks 1993
[26]Iwaniszewski 1995
[27]Zeilik 1985
[28]Zeilik 1986
[29]Milbraith 1999:8
[30]Broda 2000:233
[31]Hoskin 1996
[32]Ruggles 1993:ix
[33]Aveni 1982
[34]Heggie 1982
[35]Aveni, 1989a:xi–xiii
[36]Ruggles 2000
[37]Aveni 1981: 1-2
[38]Aveni 2003: 150
[39]McCluskey 2004
[40]Gummerman & Warburton 2005
[41]Bostwick 2006:3
[42]Bahn 1996:49
[43]McCluskey 2001
[44]Broda 2006
[45]Aldana 2007:14-15
[46]Poss 2005:97
[47]Schaefer 2006a:30
[48]Ruggles 1999: 3-9
[49]Fisher 2006
[50]Hoskin 2001:13-14.

[51] Ruggles & Saunders 1993:1-31
[52] Ruggles 2005:115-117
[53] Aveni 1986
[54] Hoskin 2001:2
[55] Ruggles & Saunders. 1993
[56] Iwaniszewski 2001
[57] Iwaniszewski 2003:7
[58] Aveni 1989:1
[59] Thom 1967: 107-117
[60] Ruggles 1999:25-29
[61] MacKie 1997
[62] MacKie 2006:362
[63] MacKie 2009
[64] Ruggles 1999:19-29
[65] Ruggles and Barclay 2000: 69-70
[66] Schaefer and Liller 1990
[67] Kintigh 1992
[68] Hoskin 2001
[69] Aveni 1989
[70] Kelley and Milone 2005:369-370
[71] Kelley and Milone 2005:367-8
[72] Milbraith 1988:70-71
[73] Aveni 2006:60-64
[74] Aveni 1979:175-183
[75] Aveni 1997:137-138
[76] Aveni 1989:5
[77] Bauer and Dearborn 1995
[78] Xu *et al.* 2000:1-7
[79] Schaefer 2006a:42-48
[80] Schaefer 2006b
[81] Iwaniszewski 2003
[82] Ruggles, 2005:112-113
[83] Ruggles 2005:423-425
[84] Scholsser 2002
[85] Meller 2004
[86] van Driel-Murray 2002
[87] T. Freeth *et al.* 2006
[88] Inscriptiones Creticae III iv 11; Isager and Skydsgaard 1992:163
[89] Williamson 1987:109-114
[90] Sofaer 2008
[91] Fountain 2005
[92] Robins & Ewing 1989
[93] Preston & Preston 2005: 115-118
[94] Science Mag, Sofaer et al. 1979: 126
[95] Cambridge U, 1982 sofaer et al.: 126
[96] Sofaer and Sinclair: 1987. UNM, ABQ
[97] Sofaer 1998: Lekson Ed, u of utah: 165
[98] Aveni 1980:40-43
[99] Šprajc and Sánchez 2013
[100] Brandt and Williamson 1979
[101] Krupp. et al. 2010: 42
[102] Ruggles 2005:89
[103] Cairns 2005
[104] Hamacher 2012
[105] Saethre 2007
[106] McCluskey 2005:78

[107]Ruggles 1999:18
[108]A.F. Aveni 1997:23-27
[109]Ruggles 1999:36-37
[110]Ruggles 2005:345-347
[111]Ruggles 2005:354-355
[112]Predicting the next bright comet http://www.space.com/spacewatch/050701_bright_comets.html, Space.com.
[113]Steel 1999
[114]Chamberlain & Young 2005:xi
[115]Turton & Ruggles 1978
[116]Aveni 1989b
[117]McCluskey 2000
[118]Salt & Boutsikas 2005
[119]Liritzis.I and Vassiliou.H (2006) Were Greek temples oriented towards aurora? *Astronomy & Geophysics*, vol.47, 2, 1.14-1.18
[120]Liritzis.I and Vassiliou.H (2006) Does sunrise day correlate with eastern orientation of Byzantine Churches during significant solar dates and Saint's day name? A preliminary study. *Byzantinische Zeitscrift* (K.G.Saur Munchen, Leipzig) 99, 2, 523-534
[121]Bauer & Dearborn 1995
[122]Urton 1981
[123]Krupp 1997a:196–9
[124]Hoskin 1999:15–6
[125]Hugh-Jones 1982:191-3
[126]Schaefer 2002
[127]Blomberg 2003, esp page 76
[128]*The Pleiades in mythology* http://www.pleiade.org/pleiades_02.html, Pleiade Associates, Bristol, United Kingdom, accessed June 7, 2012
[129]Giorgio de Santillana & Hertha von Dechend, *Hamlet's Mill*, David R Godine: Boston, 1977.
[130]Hannah 1994
[131]Krupp 1997a:252–3
[132]Dearborn, Seddon & Bauer, 1998
[133]Krupp 1988
[134]Rufinus
[135]Clive Ruggles and Michel Cotte (ed.), Heritage Sites of Astronomy and Archaeoastronomy. ICOMOS and IAU, Paris, 2010.
[136]Krupp. 1979:1
[137]Eogan 1991
[138]O'Kelly 1982:123-124
[139]Belmonte 2001
[140]Neugebauer 1980
[141]Magli 2013
[142]Spence 2000
[143]Hancock 1996:168
[144]Fairall 1999
[145]Krupp 1997b
[146]Krupp 1997a:267-269
[147]Šprajc and Sánchez 2013
[148]Parker Pearson *et al.* 2007
[149]Aveni 1997:65-66
[150]Ruggles 2005:163-165
[151]Šprajc 2015
[152]Science Magazine, Sofaer et al., 1979: 126
[153]Malville and Putnam, 1989. Johnson Books:111
[154]Sofaer 1998. Lekson Ed, U of Utah: 165.
[155]Sofaer, Marshall and Sinclair, 1989. Cambridge: 112.

[156] Julien d'Huy. 2012. Un ours dans les étoiles: recherche phylogénétique sur un mythe préhistorique. https//www.academia.edu. *Préhistoire du sud-ouest*, 20 (1), 91-106; Julien d'Huy. 2013. A Cosmic Hunt in the Berber sky https//www.academia.edu. *Les Cahiers de l'AARS*, 16, 93-106

[157] Sir Jocelyn Stephens quoted in *The Times*, 8 July 1994, 8.

[158] Pedersen 1982:269

[159] Witzel 2001

[160] Pingree 1982:554-563, esp. p. 556

[161] Gallagher 1983

[162] Pyle 1983

[163] Fell 1983

[164] Wise 2003

[165] Lesser, 1983

[166] http://www.archaeoastronomy.org/

[167] http://www.archeoastronomy.org/

[168] https://web.archive.org/web/20070927043749/http://www.arqueoastronomia.org/siac.htm

[169] http://members.ozemail.com.au/~mmichie/bulletin_oct13.pdf

[170] http://www.scoop.co.nz/stories/SC1306/S00043/modern-uses-for-ancient-maori-knowledge.htm

[171] http://www.tifr.res.in/~archaeo/

[172] http://www.thehindu.com/sci-tech/science/first-indian-record-of-supernova-found-in-kashmir/article2221426.ece

[173] //doi.org/10.1017%2FS0003598X0003252X

[174] http://adsabs.harvard.edu/abs/1986S&T....72..456A

[175] http://www.fcaglp.unlp.edu.ar/~sixto/arqueo/curso/AncientAmericasastronomy.pdf

[176] //doi.org/10.1023%2FA%3A1022971730558

[177] http://adsabs.harvard.edu/abs/2001JHAS...32....1B

[178] http://adsabs.harvard.edu/abs/1979JHAS...10....1B

[179] http://www.wam.umd.edu/~tlaloc/archastro/ae33.html

[180] http://adsabs.harvard.edu/abs/1980JHAS...11...55C

[181] //www.jstor.org/stable/971730

[182] //doi.org/10.2307%2F971730

[183] http://www.antiquityofman.com/Orion_Fairall.html

[184] https://web.archive.org/web/20080228144915/http://www.antiquityofman.com/Orion_Fairall.html

[185] http://cwva.org/wwvrunes/wwvrunes_3.html

[186] https://web.archive.org/web/20080509080053/http://cwva.org/wwvrunes/wwvrunes_3.html

[187] http://adsabs.harvard.edu/abs/2006Natur.444..587F

[188] //www.ncbi.nlm.nih.gov/pubmed/17136087

[189] //doi.org/10.1038%2Fnature05357

[190] http://cwva.org/wwvrunes/wwvrunes_2.html

[191] https://web.archive.org/web/20080511200020/http://cwva.org/wwvrunes/wwvrunes_2.html

[192] http://www.timeshighereducation.co.uk/story.asp?sectioncode=1&storycode=156832

[193] http://scholar.lib.vt.edu/ejournals/ElAnt/V2N4/hannah.html

[194] https://web.archive.org/web/20080421222540/http://scholar.lib.vt.edu/ejournals/ElAnt/V2N4/hannah.html

[195] http://www.wam.umd.edu/~tlaloc/archastro/ae9.html

[196] http://adsabs.harvard.edu/abs/1996JHAS...27...85H

[197] http://adsabs.harvard.edu/abs/1982NYASA.385..183H

[198] //doi.org/10.1111%2Fj.1749-6632.1982.tb34265.x

[199] http://www.wam.umd.edu/~tlaloc/archastro/ae18.html

[200] http://www.wam.umd.edu/~tlaloc/archastro/ae5.html

[201] http://www.antiquityofman.com/Krupp_pyramid_marketing_schemes.html

[202] http://cwva.org/ogam_rebutal/lesser_cult_arch.html

[203] http://adsabs.harvard.edu/abs/2000Arch...15...14M

[204] //doi.org/10.1017%2FS0003598X00084969

[205]//doi.org/10.2752%2F175169709x374263
[206]http://arxiv.org/pdf/physics/0703213
[207]//doi.org/10.1111%2Fj.1468-0092.2007.00296.x
[208]http://adsabs.harvard.edu/abs/1980Cent...24....1N
[209]//doi.org/10.1111%2Fj.1600-0498.1980.tb00362.x
[210]http://nora.nerc.ac.uk/9174/
[211]//doi.org/10.1017%2Fs0003598x00095624
[212]http://adsabs.harvard.edu/abs/1992Isis...83..554P
[213]//doi.org/10.1086%2F356288
[214]http://cwva.org/wwvrunes/wwvrunes_1.html
[215]https://web.archive.org/web/20080511190245/http://cwva.org/wwvrunes/wwvrunes_1.html
[216]http://www29.homepage.villanova.edu/christopher.haas/destructioni%20of%20Serapeum.html
[217]https://web.archive.org/web/20080406071455/http://www29.homepage.villanova.edu/christopher.haas/destructioni%20of%20Serapeum.html
[218]http://adsabs.harvard.edu/abs/2000ArchS..25...65R
[219]//doi.org/10.1007%2F978-1-4614-6141-8
[220]//doi.org/10.1017%2FS0003598X00066151
[221]http://www.astronomicalheritage.org/index.php?option=com_content&view=article&id=26&itemid=31
[222]//doi.org/10.1111%2Fj.1467-9655.2007.00463.x
[223]//doi.org/10.1017%2FS0003598X00114504
[224]http://adsabs.harvard.edu/abs/2002S&T...103b..38S
[225]http://adsabs.harvard.edu/abs/1990PASP..102..796S
[226]//doi.org/10.1086%2F132705
[227]http://www.britarch.ac.uk/BA/ba45/ba45feat.html#steel
[228]http://adsabs.harvard.edu/abs/2000Natur.408..320S
[229]//www.ncbi.nlm.nih.gov/pubmed/11099032
[230]//doi.org/10.1038%2F35042510
[231]https//web.archive.org
[232]//doi.org/10.1017%2FS0003598X00000053
[233]//doi.org/10.1086%2F202140
[234]http://cwva.org/ogam_rebutal/solstice_2002.html
[235]https://web.archive.org/web/20080328113612/http://ejvs.laurasianacademy.com/ejvs0703/ejvs0703d.txt
[236]http://www.ejvs.laurasianacademy.com/ejvs0703/ejvs0703d.txt
[237]http://adsabs.harvard.edu/abs/1985JHAS...16....1Z
[238]http://adsabs.harvard.edu/abs/1986JHAS...17....1Z
[239]http://assets.cambridge.org/052157/2916/sample/0521572916web.pdf
[240]http://www.cliveruggles.net/
[241]https://www.pok.polimi.it/courses/course-v1:Polimi+ArchAstr101+2016_M5/about
[242]http://www.traditionsofthesun.org/
[243]http://sunearthday.nasa.gov/2005/images/ao_poster.pdf
[244]http://siac.fcaglp.unlp.edu.ar/
[245]http://www.shastro.org.uk/
[246]http://www.wam.umd.edu/~tlaloc/archastro/ae.html
[247]http://www.shpltd.co.uk/aa.html
[248]http://www.utexas.edu/utpress/journals/jarch.html
[249]http://www.cultureandcosmos.com/
[250]http://www.shpltd.co.uk/jha.html
[251]http://aaatec.org
[252]Aaboe, Asger. "The culture of Babylonia: Babylonian mathematics, astrology, and astronomy." The Assyrian and Babylonian Empires and other States of the Near East, from the Eighth to the Sixth Centuries B.C. Eds. John Boardman, I. E. S. Edwards, N. G. L. Hammond, E. Sollberger and C. B. F. Walker. Cambridge University Press, (1991)
[253]D. Brown (2000), *Mesopotamian Planetary Astronomy-Astrology*, Styx Publications, .

[254]William P. D. Wightman (1951, 1953), *The Growth of Scientific Ideas*, Yale University Press p. 38.
[255]Index of Ancient Greek Philosophers-Scientists http://www.ics.forth.gr/~vsiris/ancient_greeks/hellinistic_period.html
[256]Seleucus of Seleucia (c. 190 BC-?) http://adsabs.harvard.edu/abs/2000eaa..bookE3998., The SAO/NASA Astrophysics Data System (ADS)
[257]Seleucus of Seleucia (ca. 190-unknown BC) http://scienceworld.wolfram.com/biography/Seleucus.html, ScienceWorld
[258]Lucio Russo, *Flussi e riflussi*, Feltrinelli, Milano, 2003, .
[259]Asger Aaboe, *Episodes from the Early History of Astronomy*, New York: Springer, 2001), pp. 62-5; Alexander Jones, "The Adaptation of Babylonian Methods in Greek Numerical Astronomy," in *The Scientific Enterprise in Antiquity and the Middle Ages*, p. 99
[260]//en.wikipedia.org/w/index.php?title=Template:Mesopotamian_myth&action=edit
[261]//en.wikipedia.org/w/index.php?title=Template:Astrology_sidebar&action=edit
[262]Holden (1996) p.1.
[263]Rochberg (1998) p.ix. See also, Neugebauer (1969) pp.29-30 https://books.google.com/books?id=JVhTtVA2zr8C.
[264]Rochberg (1998) p.x.
[265]Baigent (1994) p.71.
[266]Holden (1996) p.9.
[267]Koch-Westenholz (1995) p.16.
[268]Koch-Westenholz (1995) p.11.
[269]Koch-Westenholz (1995) p.12. Tablet source given as: *State Archives of Assyria* 8 250.
[270]Koch-Westenholz (1995) p.13.
[271]Koch-Westenholz (1995) p.19.
[272]J. G. Frazer, *The Early History of Kingship*
[273]Derek and Julia Parker, Ibid, p198, 1990
[274]http://www.tyndalehouse.com/egypt/ptolemies/chron/babylonian/chron_bab_cal_fr.htm
[275]http://www.uni-koeln.de/phil-fak/ifa/zpe/downloads/2000/130pdf/130159.pdf
[276]http://www.angelfire.com/tx/tintirbabylon/caledarintro.html
[277]https://web.archive.org/web/20091026160123/http://geocities.com/astrologymulapin/index.htm
[278]http://www.staff.science.uu.nl/~gent0113/babylon/babycal.htm
[279]Sarma (2008), *Astronomy in India*
[280]The Vedas: An Introduction to Hinduism's Sacred Texts, Roshen Dalal, p.188
[281]Highlights of Astronomy, Volume 11B: As presented at the XXIIIrd General Assembly of the IAU, 1997. Johannes Andersen Springer, 31 January 1999 – Science – 616 pages. page 721 https//books.google.com
[282]Babylon to Voyager and Beyond: A History of Planetary Astronomy. David Leverington. Cambridge University Press, 29 May 2010 – Science – 568 pages. page 41 https//books.google.com
[283]The History and Practice of Ancient Astronomy. James Evans. Oxford University Press, 1 October 1998 – History – 496 pages. Page 393 https//books.google.com
[284]Foreign Impact on Indian Life and Culture (c. 326 B.C. to C. 300 A.D.). Satyendra Nath Naskar. Abhinav Publications, 1 January 1996 – History – 253 pages. Pages 56–57 https//books.google.com
[285]"Star Maps: History, Artistry, and Cartography", p. 17, by Nick Kanas, 2012
[286]Ōhashi (1993)
[287]Hayashi (2008), *Aryabhata I*
[288]J.A.B. van Buitenen (2008)
[289]Bryant (2001), 253
[290]See A. Cunningham (1883), *A Book of Indian Eras*.
[291]
[292]Tripathi (2008)
[293]Indian Astronomy. (2013). In D. Leverington, *Encyclopedia of the history of Astronomy and Astrophysics*. Cambridge, United Kingdom: Cambridge University Press. Retrieved from
[294]Hayashi (2008), *Brahmagupta*

[295] Brahmagupta, *Brahmasphutasiddhanta* (628) (cf. al-Biruni (1030), *Indica*)

[296]

[297]

[298] Hayashi (2008), *Shripati*

[299]

[300] Joseph, 408

[301] Ramasubramanian etc. (1994)

[302] Ōhashi (2008), *Astronomical Instruments in India*

[303]

[304]

[305] Savage-Smith (1985)

[306] https://people.sc.fsu.edu/~dduke/Thursday_slides

[307] Pierre Cambon, Jean-François Jarrige. "Afghanistan, les trésors retrouvés: Collections du Musée national de Kaboul". Éditions de la Réunion des musées nationaux, 2006 – 297 pages. p269 https//books.google.com

[308] Pierre Cambon, Jean-François Jarrige. "Afghanistan, les trésors retrouvés: Collections du Musée national de Kaboul". Éditions de la Réunion des musées nationaux, 2006 – 297 pages. p269 https//books.google.com "Les influences de l'astronomie grecques sur l'astronomie indienne auraient pu commencer de se manifester plus tot qu'on ne le pensait, des l'epoque Hellenistique en fait, par l'intermediaire des colonies grecques des Greco-Bactriens et Indo-Grecs" (French) Afghanistan, les trésors retrouvés", p269. Translation: "The influence of Greek astronomy on Indian astronomy may have taken place earlier than thought, as soon as the Hellenistic period, through the agency of the Greek colonies of the Greco-Bactrians and the Indo-Greeks.

[309]

[310] Pingree, David "Astronomy and Astrology in India and Iran" *Isis*, Vol. 54, No. 2 (Jun. 1963), pp. 229–246

[311] "the Pañca-siddhāntikā ("Five Treatises"), a compendium of Greek, Egyptian, Roman and Indian astronomy. Varāhamihira's knowledge of Western astronomy was thorough. In 5 sections, his monumental work progresses through native Indian astronomy and culminates in 2 treatises on Western astronomy, showing calculations based on Greek and Alexandrian reckoning and even giving complete Ptolemaic mathematical charts and tables. Encyclopædia Britannica Source http://www.britannica.com/EBchecked/topic/623232/Varahamihira

[312]

[313] Dallal, 162

[314] King, 240

[315] Joseph, 306

[316] Sharma (1995), 8–9

[317] Baber, 82–89

[318] Raju (2001)

[319]

[320] Baber, 89–90

[321] https://archive.org/stream/hinduastronomy00brenuoft#page/n7/mode/2up

[322] https://archive.org/stream/indianeclipserep00britrich#page/n7/mode/2up

[323] http://adsabs.harvard.edu/abs/2002JHA....33..237K

[324] //doi.org/10.1353%2Fpew.2001.0045

[325] //doi.org/10.1016%2Fs0160-9327%2800%2901327-2

[326] Thurston, H., *Early Astronomy*. Springer, 1994. p.2

[327] I. Passas, K. Hasapis, *Ορφικά*. Encyclopedia Helios, 1984

[328] (Lindberg 2010, p. 86 https://books.google.com/books?id=dPUBAkIm2lUC&pg=PA86)

[329] "Afghanistan, les trésors retrouvés", p269

[330] "Les influences de l'astronomie grecques sur l'astronomie indienne auraient pu commencer de se manifester plus tot qu'on ne le pensait, des l'epoque Hellenistique en fait, par l'intermediaire des colonies grecques des Greco-Bactriens et Indo-Grecs" (French) Afghanistan, les trésors retrouvés", p269. Translation: "The influence of Greek astronomy on Indian astronomy may have taken place earlier than thought, as soon as the Hellenistic period, through the agency of the Greek colonies of the Greco-Bactrians and the Indo-Greeks.

[331]"the Pañca-siddhāntikā ("Five Treatises"), a compendium of Greek, Egyptian, Roman and Indian astronomy. Varāhamihira's knowledge of Western astronomy was thorough. In 5 sections, his monumental work progresses through native Indian astronomy and culminates in 2 treatises on Western astronomy, showing calculations based on Greek and Alexandrian reckoning and even giving complete Ptolemaic mathematical charts and tables. Encyclopædia Britannica Source http://www.britannica.com/EBchecked/topic/623232/Varahamihira

[332]":Mleccha hi yavanah tesu samyak shastram idam sthitam
Rsivat te api pujyante kim punar daivavid dvijah
-(Brhatsamhita 2.15)

[333]http://www.csit.fsu.edu/~dduke/models

[334]http://www-history.mcs.st-andrews.ac.uk/history/index.html

[335]Full version at Met Museum http://www.metmuseum.org/Collections/search-the-collections/100000870#fullscreen

[336]Ruggles, C.L.N. (2005), *Ancient Astronomy*, pages 354-355. ABC-Clio.

[337]Krupp, E.C. (1988). "Light in the Temples", in C.L.N. Ruggles: Records in Stone: Papers in Memory of Alexander Thom. CUP, 473-499.

[338]Symons, S.L., Cockcroft, R., Bettencourt, J. and Koykka, C., 2013. Ancient Egyptian Astronomy [Online database] Diagonal Star Tables http://aea.physics.mcmaster.ca/index.php/en/database/diagonal-star-tables

[339]Symons, S.L. A Star's Year: The Annual Cycle in the Ancient Egyptian Sky http://www2.le.ac.uk/Members/sls25/paperstarsyear in: Steele, J.M. (Ed.), Calendars and Years: Astronomy and Time in the Ancient World. Oxbow Books, Oxford, pp. 1-33.

[340]Marshall Clagett, *Ancient Egyptian Science, Volume 2: Calendars, clocks, and astronomy.* https://books.google.com/books?id=xKKPUpDOTKAC&pg=PA53 Philadelphia: American Philosophical Society, 1995 p53

[341]Clement of Alexandria, *Stromata*, vi. 4

[342]O Neugebauer, *Egyptian Planetary Texts*, Transactions, American Philosophical Society, Vol. 32, Part 2, 1942, Page 237.

[343]//doi.org/10.1080%2F00033790110095813

[344]Needham, Volume 3, p.171

[345]Needham, Volume 3, p.242

[346]Needham, Volume 3, p.172-3

[347]//en.wikipedia.org/w/index.php?title=Template:History_of_science_and_technology_in_China&action=edit

[348]Joseph Needham, Volume 3, p. 109

[349]Whitfield, Susan. [2004] (2004). The Silk Road: Trade, Travel, War and Faith. British Library Staff. Serindia Publications, Inc.

[350]Yabuuti, K.: Indian and Arabian Astronomy in China. In: The Silver Jubilee volume of the Zinbun-Kagaku-Kenkyusyo, Kyoto, pp. 585–603 (1954).

[351]Mukhopadhyay, S. K. (ed.): The Sardulakarnavadana. Visvabharati, Santiniketan, pp. 46–53 and p. 104 (1954).

[352]13 Vaidya, P. L. (ed.): Lalitavistara. Darbhanga, p. 103 (1958).

[353]Peng, Yoke Ho (2000). Li, Qi and Shu: An Introduction to Science and Civilization in China. Courier Dover Publications.

[354]Sun, Xiaochun and Jacob Kistemaker. (1997). *The Chinese Sky During the Han: Constellating Stars and Society*. Leiden, New York, Köln: Koninklijke Brill. Page 21-25.

[355]Kistemaker, Jacob. Sun, Xiaochun. [1997] (1997). The Chinese Sky During the Han: Constellating Stars and Society. BRILL publishing.

[356]Milone, Eugene F. Humiston Kelley, David. Exploring Ancient Skies: An Encyclopedic Survey of Archaeoastronomy. [2005] (2005).

[357]Whitfield, Susan. [2004] (2004). The Silk Road: Trade, Travel, War and Faith. British Library Staff. Serindia Publications.

[358]James E. McClellan III; Harold Dorn. *Science and Technology in World History: An Introduction https://books.google.com/books?id=aJgp94zNwNQC*. JHU Press; 14 April 2006. p. 132.

[359]Needham, Volume 3, p.411

[360]Needham, Volume 3, 411–413.
[361]Needham, Volume 3, 227.
[362]Needham, Volume 3, p.414
[363]
[364]Christopher Cullen, "Joseph Needham on Chinese Astronomy", *Past and Present*, No. 87 (May, 1980), pp. 39-53 (45)
[365]Needham, Volume 3, p.343
[366]
[367]http//books.google.com.pk
[368]
[369]Ho, Peng Yoke. (2000). *Li, Qi, and Shu: An Introduction to Science and Civilization in China*, p. 105. Mineola: Dover Publications.
[370]Needham, Volume 3, p.444
[371]Needham, Volume 3, p.444-445
[372]Needham, Volume 3, p.438-439
[373]Sivin, Nathan. "Science in Ancient China p.22-26. http://ccat.sas.upenn.edu/~nsivin/cop.pdf
[374]Needham, Volume 3, p.445
[375]Needham, Volume 3, p.447
[376]http://www.math.nus.edu.sg/aslaksen/calendar/chinese.shtml
[377]https://web.archive.org/web/20070311000648/http://www.admin.ias.edu/hssem/pingyi.html
[378]https://web.archive.org/web/20060715181848/http://www.bao.ac.cn/english/home.asp
[379]http://hua.umf.maine.edu/China/astronomy/index.html#14
[380]//en.wikipedia.org/w/index.php?title=Template:Maya_civilization&action=edit
[381]Tedlock, Barbara, Time and the Highland Maya Revised edition (1992 Page 1) "Scores of indigenous Guatemalan communities, principally those speaking the Mayan languages known as Ixil, Mam, Pokomchí and Quiché, keep the 260-day cycle and (in many cases) the ancient solar cycle as well (chapter 4)."
[382]Miles, Susanna W, "An Analysis of the Modern Middle American Calendars: A Study in Conservation." In Acculturation in the Americas. Edited by Sol Tax, p. 273. Chicago: University of Chicago Press, 1952.
[383]See entry on *Itzamna*, in Miller and Taube (1993), pp.99–100.
[384]. For details and notes on adoption among the Mayanist community, see Kettunen & Hemke (2005), p. 5
[385]Tedlock (1992), p. 1
[386]"Mythological" in the sense that when the Long Count was first devised sometime in the Mid- to Late Preclassic, long after this date; see e.g. Miller and Taube (1993, p.50).
[387]Voss (2006, p.138)
[388]See separate brief Wikipedia article Lords of the Night
[389]Kettunen and Helmke (2011), Different Tzolk'in date of name, pp.58–59
[390]Classic-era reconstructions are as per Kettunen and Helmke (2005), pp.45–46..
[391]Kettunen and Helmke (2005), pp.47–48
[392]These names come from de Landa's description of the calendar and they are commonly used by Mayanists, but the Classic Maya did not use these actual names for the day signs. The original names are unknown. See
[393]Zero Pop actually fell on the same day as the solstice on 12/27/–575, 12/27/–574, 12/27/–573 and 12/26/–572 (astronomical year numbering, Universal Time), if you don't account for the fact that the Maya region is in roughly time zone UT–6. See IMCCE seasons http://www.imcce.fr/en/grandpublic/temps/saisons.php.
[394]Boot (2002), pp.111–114.
[395]For further details, see Thompson 1966: 123–124
[396]Thompson 1966: 124
[397]For a thorough treatment of the Year Bearers, see Tedlock 1992: 89–90; 99–104 and Thompson 1966
[398]See Coe 1965
[399]Tedlock 1992: 92

[400]Miles, Susanna W, "An Analysis of the Modern Middle American Calendars: A Study in Conservation." In Acculturation in the Americas. Edited by Sol Tax, pp. 273–84. Chicago: University of Chicago Press, 1952.
[401]As quoted in *USA Today* (MacDonald 2007).
[402]Thompson, J. Eric S. Maya Hieroglyphic Writing, 1950 Page 236
[403]Teeple 1931:53
[404]Thompson Maya Hieroglyphic Writing 1950:240
[405]Linden 1996:343–356.
[406]Schele, Grube, Fahsen 1992
[407]Teeple 1931:67
[408]Grofe, Michael John 2007 The Serpent Series: Precession in the Maya Dresden Codex page 55 p.206
[409]Maya Hieroglyphic Writing 1961 pp. 212–217
[410]Decipherment of Maya Script , David Kelley 1973 pp.56—57
[411]Star Gods of the Maya Susan Milbrath 1999, University of Texas Press
[412]Coe, William R. 'TIKAL a handbook of the ancient Maya Ruins' The University Museum of the University of Pennsylvalia, Philadelphia, Pa. 1967 p. 114
[413]Roys 1967: 132, 184–185
[414]//www.worldcat.org/oclc/45195586
[415]http://www.mesoweb.com/resources/vocabulary/Vocabulary.pdf
[416]//www.worldcat.org/issn/0011-3204
[417]//www.worldcat.org/oclc/62217742
[418]//doi.org/10.1086%2F202782
[419]//www.worldcat.org/oclc/15895415
[420]//www.worldcat.org/oclc/26605966
[421]//www.worldcat.org/oclc/50676955
[422]//www.worldcat.org/oclc/150172
[423]//www.worldcat.org/oclc/11641566
[424]http://www.mesoweb.com/resources/handbook/
[425]https://web.archive.org/web/20080316220726/http://www.usatoday.com/tech/science/2007-03-27-maya-2012_N.htm
[426]//www.worldcat.org/issn/0734-7456
[427]https://www.usatoday.com/tech/science/2007-03-27-maya-2012_N.htm
[428]//www.worldcat.org/oclc/40848420
[429]//www.worldcat.org/oclc/27667317
[430]//www.worldcat.org/oclc/59432784
[431]//www.worldcat.org/oclc/145324300
[432]//www.worldcat.org/oclc/7653289
[433]http://www.mesoweb.com/publications/CAA/CAA02.pdf
[434]//www.worldcat.org/oclc/11467786
[435]http://www.gutenberg.org/etext/18973/
[436]//www.worldcat.org/oclc/14963920
[437]//www.worldcat.org/oclc/275252
[438]//www.worldcat.org/oclc/625693
[439]//www.worldcat.org/oclc/71165439
[440]http://research.famsi.org/date_mayaLC.php
[441]http://www.teoria.com/jra/maya/index.html
[442]http://www.manana.cz/index_en.html
[443]//en.wikipedia.org/w/index.php?title=Template:Aztecsmall&action=edit
[444]The Mexica Calendar and the Cronography, Rafael Tena. INAH-CONACULTA. 2008
[445]Crónica Mexicayotl, Fernando Alvarado Tezozomoc p 36
[446]https://web.archive.org/web/20080625221516/http://www.famsi.org/research/aguilar/Aztec_Art_Bib.pdf
[447]http://www.famsi.org/research/aguilar/Aztec_Art_Bib.pdf
[448]//www.worldcat.org/oclc/50090230
[449]//www.worldcat.org/oclc/45144264

[450]http://www.doaks.org/Native/trad07.pdf
[451]//www.worldcat.org/oclc/34354931
[452]//www.worldcat.org/oclc/40939882
[453]//www.worldcat.org/oclc/71632174
[454]//www.worldcat.org/oclc/54014738
[455]//www.worldcat.org/oclc/29708907
[456]//www.worldcat.org/oclc/44167649
[457]https://web.archive.org/web/20080515150042/http://www.arqueomex.com/S2N3nLENGUA70.html
[458]//www.worldcat.org/issn/0188-8218
[459]http://arqueomex.com/S2N3nLENGUA70.html
[460]https://web.archive.org/web/20060103090214/http://www.arqueomex.com/S8N4ICONKlein55.pdf
[461]http://arqueomex.com/S8N4ICONKlein55.pdf
[462]//www.worldcat.org/oclc/181727
[463]http://www.dartmouth.edu/~izapa/M-1.pdf
[464]http://adsabs.harvard.edu/abs/1973Sci...181..939M
[465]//www.ncbi.nlm.nih.gov/pubmed/17835843
[466]//doi.org/10.1126%2Fscience.181.4103.939
[467]https://web.archive.org/web/20080528152046/http://www.dartmouth.edu/~izapa/M-1.pdf
[468]//www.worldcat.org/oclc/27667317
[469]//www.worldcat.org/oclc/37909790
[470]//www.worldcat.org/oclc/276351
[471]//www.worldcat.org/oclc/48579073
[472]//www.worldcat.org/oclc/43337963
[473]http://sites.estvideo.net/malinal/nahuatl.page.html
[474]//www.worldcat.org/oclc/11261299
[475]http://www.ancient.eu/article/896/
[476]http://www.mna.inah.gob.mx/
[477]http://www.azteccalendar.com/
[478]Saliba (1999).
[479]Dallal (1999), pg. 162
[480]This book is not related to al-Khwarizmi's *Zij al-Sindh*. On *zijes* see E. S. Kennedy, "A Survey of Islamic Astronomical Tables".
[481]Dallal (1999), pg. 163
[482]Dallal (1999), pg. 164
[483]Dallal (1999), pg. 171
[484]
[485]Ho, Peng Yoke. (2000). *Li, Qi, and Shu: An Introduction to Science and Civilization in China*, p. 105. Mineola: Dover Publications.
[486]John Morris Roberts, *The History of the World*, pp. 264–74, Oxford University Press,
[487]http://museodeco.com/t-prodReplogle.aspx#q9
[488]Richard Nelson Frye: Golden Age of Persia. p. 163.
[489]https://web.archive.org/web/20071031023004/http://www.soas.ac.uk/gallery/Previous/IslamicPatronage/19.html
[490]http://www.astrolabes.org/pages/saphea.htm
[491]Tuncer Oren (2001). "Advances in Computer and Information Sciences: From Abacus to Holonic Agents", *Turk J Elec Engin* **9** (1): 63–70 [64].
[492]David A. King, "Islamic Astronomy," pp. 168–9.
[493]David A. King, "Islamic Astronomy," pp. 167–8.
[494]https://web.archive.org/web/20080102034220/http://www.chowk.com/articles/9489
[495]http://www.chowk.com/articles/9489
[496]https://web.archive.org/web/20110101211150/http://faculty.kfupm.edu.sa/PHYS/alshukri/PHYS215/Islamic_astronomy.htm
[497]http://adsabs.harvard.edu/abs/1986SciAm.254...74G
[498]//doi.org/10.1038%2Fscientificamerican0486-74

[499] http://faculty.kfupm.edu.sa/PHYS/alshukri/PHYS215/Islamic_astronomy.htm
[500] https://web.archive.org/web/20080218171021/http://www.history-science-technology.com/Articles/articles%2071.htm
[501] http://www.history-science-technology.com/Articles/articles%2071.htm
[502] //doi.org/10.1080%2F00033798500200141
[503] http://home.swipnet.se/islam/articles/HistoryofSciences.htm
[504] https://web.archive.org/web/20071225091836/http://home.swipnet.se/islam/articles/HistoryofSciences.htm
[505] //doi.org/10.1086%2F348036
[506] //doi.org/10.1086%2F349146
[507] //www.jstor.org/stable/595221
[508] //doi.org/10.2307%2F595221
[509] //doi.org/10.1086%2F349363
[510] //www.jstor.org/stable/1005726
[511] //doi.org/10.2307%2F1005726
[512] //doi.org/10.1086%2F371617
[513] //doi.org/10.1086%2F349558
[514] //doi.org/10.1086%2F353360
[515] //doi.org/10.1080%2F03085699708592859
[516] http://adsabs.harvard.edu/abs/2002JHA....33..237K
[517] //doi.org/10.1111%2Fj.1600-0498.2003.450117.x
[518] //doi.org/10.1080%2F00033790110095813
[519] //www.jstor.org/stable/2851429
[520] //doi.org/10.2307%2F2851429
[521] //doi.org/10.1017%2Fs0269889701000060
[522] http://adsabs.harvard.edu/abs/2001Osir...16...49R
[523] //doi.org/10.1086%2F649338
[524] http://www.npr.org/templates/story/story.php?storyId=885213
[525] https://web.archive.org/web/20080101212125/http://www.npr.org/templates/story/story.php?storyId=885213
[526] //doi.org/10.1017%2FS0957423907000355
[527] //www.jstor.org/stable/2709773
[528] //doi.org/10.2307%2F2709773
[529] //doi.org/10.1086%2F352344
[530] //www.jstor.org/stable/601763
[531] //doi.org/10.2307%2F601763
[532] http://adsabs.harvard.edu/abs/1994JHA....25..115S
[533] http://www.columbia.edu/~gas1/project/visions/case1/sci.1.html
[534] https://web.archive.org/web/20080115070924/http://www.columbia.edu/~gas1/project/visions/case1/sci.1.html
[535] //doi.org/10.1162%2F106361400753373713
[536] https://www.youtube.com/watch?v=IkL5Klh9WlY
[537] https://web.archive.org/web/20110830055134/http://sunucu1.tug.tubitak.gov.tr/index.html?en
[538] https://web.archive.org/web/20051218025632/http://faculty.kfupm.edu.sa/phys/alshukri/PHYS215/Islamic%20astronomy.htm
[539] http://www.auass.com/
[540] https://web.archive.org/web/20070707114137/http://www.kacst.edu.sa/eng/inst/agri/dept4.php
[541] http://www.hps.cam.ac.uk/starry/isaslabecalen.html
[542] http://www.ianridpath.com/startales/alsufi.htm
[543] In Greek, αἰτιολογητός means "explained, justified" (from αἰτιολογῶ "I explain, I justify").
[544] Here G.V. is a siglum for "Generositas Vestra", see
[545] In his *Astronomia nova*, Kepler presented only a proof that Mars' orbit is elliptical. Evidence that the other known planets' orbits are elliptical was presented later. See: Johannes Kepler, *Astronomia nova* ... (1609), p. 285. https://archive.org/stream/Astronomianovaa00Kepl#page/284/mode/2up After having rejected circular and oval orbits, Kepler concluded that Mars'

orbit must be elliptical. From the top of page 285: *"Ergo ellipsis est Planetæ iter; ... "* (Thus, an ellipse is the planet's [i.e., Mars'] path; ...) Later on the same page: *" ... ut sequenti capite patescet: ubi simul etiam demonstrabitur, nullam Planetæ relinqui figuram Orbitæ, præterquam perfecte ellipticam; ... "* (... as will be revealed in the next chapter: where it will also then be proved that any figure of the planet's orbit must be relinquished, except a perfect ellipse; ...) And then: *"Caput LIX. Demonstratio, quod orbita Martis, ... , fiat perfecta ellipsis: ... "* (Chapter 59. Proof that Mars' orbit, ... , be a perfect ellipse: ...) The geometric proof that Mars' orbit is an ellipse appears as Protheorema XI on pages 289-290.
Kepler stated that all planets travel in elliptical orbits having the Sun at one focus in: Johannes Kepler, *Epitome Astronomiae Copernicanae* [Summary of Copernican Astronomy] (Linz ("Lentiis ad Danubium"), (Austria): Johann Planck, 1622), book 5, part 1, III. De Figura Orbitæ (III. On the figure [i.e., shape] of orbits), pages 658-665. https://books.google.com/books?id=wa2SE_6ZL7YC&pg=PA658#v=onepage&q&f=false From p. 658: *"Ellipsin fieri orbitam planetæ ... "* (Of an ellipse is made a planet's orbit ...). From p. 659: *" ... Sole (Foco altero huius ellipsis) ... "* (... the Sun (the other focus of this ellipse) ...).

546 In his *Astronomia nova* ... (1609), Kepler did not present his second law in its modern form. He did that only in his *Epitome* of 1621. Furthermore, in 1609, he presented his second law in two different forms, which scholars call the "distance law" and the "area law". • His "distance law" is presented in: *"Caput XXXII. Virtutem quam Planetam movet in circulum attenuari cum discessu a fonte."* (Chapter 32. The force that moves a planet circularly weakens with distance from the source.) See: Johannes Kepler, *Astronomia nova* ... (1609), pp. 165-167. https://archive.org/stream/Astronomianovaa00Kepl#page/164/mode/2up On page 167 https://archive.org/stream/Astronomianovaa00Kepl#page/166/mode/2up, Kepler states: *" ... , quanto longior est αδ quam αε, tanto diutius moratur Planeta in certo aliquo arcui excentrici apud δ, quam in æquali arcu excentrici apud ε."* (... , as αδ is longer than αε, so much longer will a planet remain on a certain arc of the eccentric near δ than on an equal arc of the eccentric near ε.) That is, the farther a planet is from the Sun (at the point α), the slower it moves along its orbit, so a radius from the Sun to a planet passes through equal areas in equal times. However, as Kepler presented it, his argument is accurate only for circles, not ellipses. • His "area law" is presented in: *"Caput LIX. Demonstratio, quod orbita Martis, ... , fiat perfecta ellipsis: ... "* (Chapter 59. Proof that Mars' orbit, ... , is a perfect ellipse: ...), Protheorema XIV and XV, pp. 291-295. https://archive.org/stream/Astronomianovaa00Kepl#page/284/mode/2up On the top p. 294, it reads: *"Arcum ellipseos, cujus moras metitur area AKN, debere terminari in LK, ut sit AM."* (The arc of the ellipse, of which the duration is delimited [i.e., measured] by the area AKM, should be terminated in LK, so that it [i.e., the arc] is AM.) In other words, the time that Mars requires to move along an arc AM of its elliptical orbit is measured by the area of the segment AMN of the ellipse (where N is the position of the Sun), which in turn is proportional to the section AKN of the circle that encircles the ellipse and that is tangent to it. Therefore, the area AMN that is swept out by a radius from the Sun to Mars as Mars moves along an arc AM of its elliptical orbit is proportional to the time that Mars requires to move along that arc. Thus, a radius from the Sun to Mars sweeps out equal areas in equal times. In 1621, Kepler restated his second law for any planet: Johannes Kepler, *Epitome Astronomiae Copernicanae* [Summary of Copernican Astronomy] (Linz ("Lentiis ad Danubium"), (Austria): Johann Planck, 1622), book 5, page 668 https://books.google.com/books?id=wa2SE_6ZL7YC&pg=PA668#v=onepage&q&f=true. From page 668: *"Dictum quidem est in superioribus, divisa orbita in particulas minutissimas æquales: accrescete iis moras planetæ per eas, in proportione intervallorum inter eas & Solem."* (It has been said above that, if the orbit of the planet is divided into the smallest equal parts, the times of the planet in them increase in the ratio of the distances between them and the sun.) That is, a planet's speed along its orbit is inversely proportional to its distance from the Sun. (The remainder of the paragraph makes clear that Kepler was referring to what is now called angular velocity.)

547 Johannes Kepler, *Harmonices Mundi* [The Harmony of the World] (Linz, (Austria): Johann Planck, 1619), p. 189. https://books.google.com/books?id=ZLlCAAAAcAAJ&pg=PA189#v=onepage&q&f=false From the bottom of p. 189: *"Sed res est certissima exactissimaque quod* proportio qua est inter binorum quorumcunque Planetarum tempora periodica, sit præcise

sesquialtera proportionis *mediarum distantiarum, ... "* (But it is absolutely certain and exact that the *proportion between the periodic times of any two planets is precisely the sesquialternate proportion* [i.e., the ratio of 3:2] of their mean distances, ... ")
An English translation of Kepler's *Harmonices Mundi* is available as: Johannes Kepler with E.J. Aiton, A.M. Duncan, and J.V. Field, trans., *The Harmony of the World* (Philadelphia, Pennsylvania: American Philosophical Society, 1997); see especially p. 411 https://books.google.com/books?id=rEkLAAAAIAAJ&pg=PA411#v=onepage&q&f=false.

[548] Johannes Kepler, *Astronomia nova* ... (1609), p. 5 of the *Introductio in hoc opus* (Introduction to this work). From page 5: https://archive.org/stream/Astronomianovaa00Kepl#page/n24/mode/1up *"Orbis virtutis tractoriæ, quæ est in Luna, porrigitur utque ad Terras, & prolectat aquas sub Zonam Torridam, ... Celeriter vero Luna verticem transvolante, cum aquæ tam celeriter sequi non possint, fluxus quidem fit Oceani sub Torrida in Occidentem, ... "* (The sphere of the lifting power, which is [centered] in the moon, is extended as far as to the earth and attracts the waters under the torrid zone, ... However the moon flies swiftly across the zenith ; because the waters cannot follow so quickly, the tide of the ocean under the torrid [zone] is indeed made to the west, ...)

[549] http://www.e-rara.ch/zut/content/titleinfo/162514

[550] https://web.archive.org/web/20060920211226/http://www.library.usyd.edu.au/libraries/rare/modernity/kepler4.html

[551] Lawrence Nolan (ed.), *The Cambridge Descartes Lexicon*, Cambridge University Press, 2016, "Inertia."

[552] https://books.google.gr/books?id=KUE1AAAAcAAJ&vq=inertia&hl=el&source=gbs_navlinks_s&redir_esc=y

[553] The [[Quito Astronomical Observatory http://oaq.epn.edu.ec/] is managed by National Polytechnic School, EPN, official web site.]

[554] http://lot.astro.utoronto.ca/

[555] The ESO 100-m OWL optical telescope concept http://www.eso.org/projects/owl

Article Sources and Contributors

The sources listed for each article provide more detailed licensing information including the copyright status, the copyright owner, and the license conditions.

Archaeoastronomy *Source:* https://en.wikipedia.org/w/index.php?oldid=806848472 *License:* Creative Commons Attribution-Share Alike 3.0 *Contributors:* 564dude, 83d40m, A. Parrot, Aarghdvaark, Advanceddeepspacepropeller, Agyle, All Is One, Alonso de Mendoza, Alunsalt, Anrnusna, BCtl, Bender235, Bernietaylor, Bheldthor, Bobamnertiopsis, Bpenprase, CJLL Wright, CXCV, Canterbury Tail, CarloMartinelli, Casliber, Catsmoke, Chris the speller, Citation bot 1, Ckatz, ClovisPt, ClueBot NG, ColJHF, Colonies Chris, CommonsDelinker, Constructive editor, Crzer07, DadaNeem, Darth Tacker, Date delinker, Dbachmann, Dcirovic, Dewritech, Dhamacher, Dhebold, Dodshe, Doug Weller, Drmies, Dsp13, DuaneHamacher, Edward, Eldizzino, Enric Naval, Euanmackie, Evensteven, Faolin42, Florian Blaschke, Furkaocean, Furkhaocean, Ginsuloft, Graham87, Graywords, Guillaume2303, HRIN, Headbomb, Hillbillyholiday, Homonihilis, Horologium, Hovgiv, Icairns, Ikey Ike, JaGa, JamesAM, Jezhotwells, Justlettersandnumbers, K6ka, KConWiki, Kadros∼enwiki, Keri, Kintetsubuffalo, Ktr101, Little sawyer, LlywelynII, Mar11, Marcog11po1, Mardhil, Mark Ironie, Matthew Yeager, Maurice Carbonaro, Mboverload, Mejor Los Indios, Michael Hardy, Mike s, Mitch Ames, Modest Genius, Myasuda, Natg 19, Netherzone, Neutral current, Niceguyedc, Nishkid64, No Swan So Fine, Oddacon, Originalwana, P. S. Burton, PatHadley, Paulscrawl, Petrb, PleaseStand, PlyrStar93, Portillo, RJHall, RabidDeity, Radioactive afikomen, Randy Kryn, Rfassbind, Rich Farmbrough, Rjwilmsi, Roberto25, Ruud Koot, SLClemens, SaturnCat, Schrauwers, Second Quantization, ShelfSkewed, Simonm223, Skyerise, Skysmith, Snokite, SoxBot III, Sprajc, SteveMcCluskey, Sun Creator, Sundial12345, Sundiall, TAnthony, Terrymacro, Theroadislong, Thoughtfortheday, Tom.Reding, Toyotsu, Trappist the monk, Ufanzis, Vanisaac, VanishedUser sdu9aya9fasdsopa, Vsmith, Wandalstouring, Wargo, Wavelength, Werieth, Wikid77, Woohookitty, Xpaulk, Y-barton, 100 anonymous edits 1

Babylonian astronomy *Source:* https://en.wikipedia.org/w/index.php?oldid=796680670 *License:* Creative Commons Attribution-Share Alike 3.0 *Contributors:* Al-Andalusi, Alansohn, Allstarecho, Appicharlask, BCtl, Baldeagle69, Bear-rings, Bellerophon5685, Bender235, BinaryFlower, Brandmeister, CLCStudent, Caeruleancentaur, Calmer Waters, Chrissymad, ClueBot NG, D0762, Danger, DarTar, Dbachmann, Dcirovic, Doug Weller, Ed Poor, ExperiencedArticleFixer, FlavrSavr, FlyingToaster, Frze, Geologist258, GermanJoe, Giftlite, Gunblade∼enwiki, HTML2011, Hakkari, Harizotoh9, Headbomb, Hmains, Hmainsbot1, Hurleywurley, IansAwesomePizza, Iridescent, Irānshahr, Italia2006, Izzedine, J04n, Jagged 85, JaidynM, Jim.henderson, John D. Croft, KConWiki, Kaiba, Ketsuekigata, Khazar2, Khirurg, Kishorekumar 62, Lagoset, LarryJeff, Macaddct1984, Maestlin, Materialscientist, Maurog, Michael Hardy, Nasa-verve, Newzild, Non-dropframe, Omnipaedista, Orphan Wiki, Ottershrew, Pearle, Profanatas, R'n'B, Rich Farmbrough, Rjwilmsi, SJ Defender, SMC, Sean.hoyland, Serendipodous, Sionus, Srleffler, Staszek Lem, SteveMcCluskey, Stvnkir, Super-Magician, Syncategoremata, The TriZ, Thirdright, Tim1357, Tolly4bolly, Tom.Reding, Vishnava, Vortulo, WadeSimMiser, Xorichaxoxo, Zarex, Zasimon, 87 anonymous edits 45

Babylonian astrology *Source:* https://en.wikipedia.org/w/index.php?oldid=782756357 *License:* Creative Commons Attribution-Share Alike 3.0 *Contributors:* Akerans, AnonMoos, Appraiser, Astrologist, BD2412, Blibintheblob, Bota47, Chris Brennan, Closedmouth, ClueBot NG, Colonies Chris, Crown Prince, Czj, DarkAudit, Dbachmann, Doug Weller, Dream of Nyx, Epbr123, Felizdenovo, FreedomAtLast, Frietjes, Gadget850, HiDrNick, Hung Mieu, IansAwesomePizza, Itc editor2, Izzedine, Jagged 85, Jessez70, John Halloran, Kidwiki91, Leszek Jańczuk, Lizia7, Maestlin, Magioladitis, Materialscientist, Maurice Carbonaro, Mitch Ames, NJMauthor, Nman360, Novangelis, Onebravemonkey, Palindromedairy, Peterstrempel, Piano non troppo, Rgbea, Rich Farmbrough, Richerman, Rjm at sleepers, Rjwilmsi, Rudimae, Rursus, Samuel Grant, Searchme, SirEbenezer, Some jerk on the Internet, Sumerophile, Syncategoremata, Synchronism, Terrymacro, UtDicitur, Valentinejoesmith, Vsmith, Woohookitty, Wykis, Zachariel, 40 anonymous edits 54

Babylonian calendar *Source:* https://en.wikipedia.org/w/index.php?oldid=782591535 *License:* Creative Commons Attribution-Share Alike 3.0 *Contributors:* 2ct7, 4pq1injbok, Abosaleh911, AstroLynx, Azkar, Belovedfreak, Bgwhite, Bravehearted, Bryan Derksen, Caerwine, Cal Enda Gar, ClueBot NG, Cyrius, Dave314159, Dbachmann, Doronve, Doug Weller, Dream of Nyx, Dthomsen8, Enthusiast01, FKmailliW, FilipeS, Fratrep, Guy1890, Hmains, IansAwesomePizza, Icairns, Imran, Iridescent, Irānshahr, Italus, Izzedine, J.delanoy, Jaeger123, Jayjg, Jimhoward72, Joe Kress, John J. Bulten, John of Reading, Karl Palmen, Kimse, Lindert, Little Mountain 5, LlywelynII, Marc Venot, Martin Thomps, Menchi, Michael Hardy, Misibacsi, Modernist, MusikAnimal, Navdar, Ogress, Penguins53, Peter Karlsen, Pseudomonas, Puthoni, RScheiber, Rich Farmbrough, Rktect, Rmhermen, ShelfSkewed, SiobhanHansa, SoHome, SplingyRanger, StAnselm, Summer Song, Template namespace initialisation script, Terabyte∼enwiki, V-VA, Vinay Jha, VoABot II, Wikalliz, Yamara, 60 anonymous edits 61

Indian astronomy *Source:* https://en.wikipedia.org/w/index.php?oldid=803784887 *License:* Creative Commons Attribution-Share Alike 3.0 *Contributors:* 89sec, Abecedare, Ajaysabarish, Aldis90, Ali944rana, Anachronist, Astropeeyush, BD2412, Bagatelle, Belasd, Bender235, Bgwhite, Bladesmulti, Blaxthos, Brandmeister, Capitals00, Citation bot 1, ClueBot NG, Dbachmann, Dcirovic, Devanampriya, Diannaa, Donner60, Doug Weller, Dr.K., Dsp13, Ekabhishek, Gaius Cornelius, Gangamohannathi, Gopik19, Hdt83, Hell in a Bucket, Hillbillyholiday, Himanish Dutta, Hmains, Italia2006, Iṣṭa Devatā, J04n, JSR, Jagged 85, JayC, Jc3s5h, John of Reading, KConWiki, Kaushal42, Kautilya3, Khirurg, Kishorekumar 62, Kkm010, Klbrain, Krishnachandranvn, LilHelpa, LouisAragon, M-le-mot-dit, MB, Malcolma, Martarius, Materialscientist, Mercurywoodrose, Mitch Ames, Morutyta, Mrkks, MusikAnimal, Nasa-verve, Neeraj Bhakta, NicatronTg, Ninney, Ohconfucius, Orenburg1, Oshwah, Physchim62, Piguy101, Qwertyus, RJHall, Raghvendra99674010, Raguks, Rbarnett93, Redtigerxyz, Rich Farmbrough, Rjwilmsi, Rror, Rudrasharman, Rursus, Rzvas, Saltine, Second Quantization, Shreevatsa, Sideways713, SteveMcCluskey, The Rambling Man, TheDJ, Timmyshin, Tom.Reding, Tony1, Trappist the monk, Tulocci, TwistedByFate, Useight, Vanished user zm34pq51mz, Vedicsciences, WOSlinker, Widr, William M. Connolley, Wujastyk, Zerokitsune, தென்காசி சுப்பிரமணியன், 116 anonymous edits 64

Ancient Greek astronomy *Source:* https://en.wikipedia.org/w/index.php?oldid=804067175 *License:* Creative Commons Attribution-Share Alike 3.0 *Contributors:* ***Ria777, 72, Abbott75, Aldaron, Annielogue, Arthur Holland, Astrologist, BD2412, Bender235, Bgpaulus, BlytheG, Bobo192, BoogaLouie, BrandonTheBro123, C.Fred, CAPTAIN RAJU, Caiaffa, Cenarium, Chanheigeorge, Chrysa D, Clarince63, ClueBot NG, Conone, Czj, DMacks, Daatass, Dardanos, DavidJac, Davidiad, Dbachmann, Dcirovic, Denisarona, Deor, Doug Weller, Drosdaf, Durova, Edison, Editor2020, Epbr123, Erpingham, Extramural, Extraordinary, Falcorian, Fieldday-sunday, Flamarande, FoCuSandLeArN, Fried Gold, Fuhghettaboutit, Genesai, Geologist258, Gguidelas, Giftlite, Gilliam, Gladtobeherenow, Guardianofhistory, Gun Powder Ma, Gwernol, Hayman30, Hmains, Houutata, Hugo999, Ian.thomson, Immunize, Invmog, Ishmayl, J.delanoy, Jagged 85, Jhenderson777, JmA, Jocularjabes, Kazvorpal, Khirurg, Kxx, Laurapr, Leinad-Z, Lemnaminor, Limttado, Llywrch, MadGuy7023, Maestlin, Manuela Revello, Mark9296, Martarius, Materialscientist, Maurog, Michael Hardy, Minimac, Mitch Ames, Morutyta, Nasa-verve, NawlinWiki, Nihiltres, Noym, Onebravemonkey, Oota∼enwiki, Papa November, PasabaPorAqui, Pasicles, Paul August, Per Honor et Gloria, Piano non troppo, Porterjoh, Postdlf, Ppole, R'n'B, Rajaram Sarangapani, RexNL, Rich Farmbrough, Rjwilmsi, Robin Hood∼enwiki, Rror, Sardanaphalus, Sergio Macías, Serols, Shirulashem, Shogartu, Singinglemon∼enwiki, Smmalut, StaticGull, SteveMcCluskey, Teammoto, The elephant, Tom.Reding, Top.Squark, Tot12, Twthmoses, Ummairsaeed, Vigyani, Vrenator, WadeSimMiser, Widr, WikHead, Will Beback, Wknight94, Wujastyk, YuviPanda, Zbayz, 199 anonymous edits 78

Egyptian astronomy *Source:* https://en.wikipedia.org/w/index.php?oldid=805538289 *License:* Creative Commons Attribution-Share Alike 3.0 *Contributors:* A. Parrot, Archeologo, Bender235, Calmer Waters, Church, Citation bot 1, ClueBot NG, Colonies Chris, CommonsDelinker, David J Wilson, Dbachmann, Dcirovic, DemocraticLuntz, Deneb in Cygnus, Dialectric, DivermanAU, Dobby102, Donner60, Doug Weller, Epicgenius, Franco3450, Glane23, Golfcourseairhorn, Guy1890, Hmains, Hobartimus, Horst-schlaemma, Ionutzmovie, Italia2006, Jagged 85, Khirurg, Khruner, KingQuail, Leoboudv, Llywrch, Marek69, Mitch Ames, Mo-Slimy∼enwiki, Mr Mandudeguy, Nasa-verve, NewEnglandYankee, Onebravemonkey, Oskar71, PericlesofAthens, Puffin, QuentinObis, Rich Farmbrough, RickLive, Roberto Belisário, Sadads, SamuelTheGhost, Sardanaphalus, Seeleschneider, Shinpah1, Singinglemon∼enwiki, Soupforone, SteveMcCluskey, Syncategoremata, ThePlatypusofDoom, Trappist the monk, Trulybaka, Vanguardhighschool, VanishedUser sdu9aya9fasdsopa, Vlādis Mānisqā, WOSlinker, Wavelength, Wikipelli, Wtmitchell, Xezbeth, Y-barton, 63 anonymous edits 89

Chinese astronomy *Source:* https://en.wikipedia.org/w/index.php?oldid=796108845 *License:* Creative Commons Attribution-Share Alike 3.0 *Contributors:* 468SM, 83d40m, Amorymeltzer, AngChenrui, Apainterofdreams, Aranel, Arcturus, BD2412, BW95, Balthazarduju, Barticus88, Bejnar, Beland, Bender235, Benjwong, Binksternet, Black Knight takes White Queen, Blethering Scot, BradBeattie, Brandmeister, CIS, CRGreathouse, CanisRufus, Casliber, Cattus, Chiswick Chap, ChrisCork, ClueBot NG, Copyeditor42, Cyfal, DOSGuy, DancingPhilosopher, Dbachmann, Derek R Bullamore, Dobs∼enwiki, DrKay, Drpickem, Dsp13, Einstein Li 37, Eiorgiomugini, FayssalF, Freefall333, Galorr, Gareth Griffith-Jones, GeneralCheese, GoShow, Grimhelm, Haida19, HeartofaDog, Hmains, Hmainsbot1, Icairns, Ideogram, Imminent77, Incrassate, Iridescent, J04n, J8079s, Jagged 85, Jmencisom, Jncraton, John, Jonathandude3000, Jusdafax, Katherine Tredwell, Keahapana, Keviniga, Khanate General, Khirurg, Krishnachandranvn, Linas, Ling Kah Jai, LiuSun, LlywelynII, Lotje, Lou2shi1, Luk, Maestlin, Matthew Yeager, Melaen, Midas02, Milton Stanley, Missvain, Mxn, Namasteia, Nasa-verve, Nishidani, Njaelkies Lea, Nlu, Nyq, O.Koslowski, Orphan Wiki, PericlesofAthens, Petrzak, Pfctdayelise, Planktonamore, Ptbotgourou, Quadpus, R'n'B, RSStockdale, Raghvendra99674010, RaseaC, Remember the dot, Rurik the Varangian, RyanGerbil10, S200401064, Sdlitvin, ShakespearePotter24601, Shellwood, Skyfiler, Soupmaker1975, StAnselm, SteveMcCluskey, Super48paul, Swerdnaneb, Syrthiss, Tamorlan, Tassedethe, Techfast50, Thecheesykid, Topbanana, Traveletti, Underbar dk, Utcursch, VoABot II, War wizard90, Who, Wikitza, Woohookitty, Xingguan, Zygimantus, 124 anonymous edits 96

Maya calendar *Source:* https://en.wikipedia.org/w/index.php?oldid=805046975 *License:* Creative Commons Attribution-Share Alike 3.0 *Contributors:* 3edw, Agyle, Algoquenoesteusado, Allanth, Allthefoxes, AlphaBetaGamma01, Altenmann, Annalysia420, Asdjhfaindsvhnjbxugvyr, AstroLynx, BethNaught, BleffeIN, Britmax, BroVic, CAPTAIN RAJU, ClueBot NG, Craic Den, Crystallizedcarbon, Dawnseeker2000, Dcirovic, Donner60, Doug Weller, Editor2020, Emeraldcityserendipity, EuroCarGT, Excirial, Flyer22 Reborn, Formuse, GhicksAA, Gilliam, Ginsuloft, Golu sasha, GreenC, Grr, Hairy Dude, Hmains, Howard McCay, Howcheng, I dream of horses, JESUS WHY, JGodwisdom, Jackmcbarn, Jasonasosa, Jennifer.sierrawifi, Jim1138, Jodosma, John of Reading, Jonathansuh, João Pedro Corrêa Eboli, K6ka, Kalendis, Karl Palmen, Klbrain, Kvantikos afros, KylieTastic, Lacambalam,

Loveifmoney, MadCowpoke, Magioladitis, Manuel chuenquitze, Manul, Marek69, Mark Arsten, MarnetteD, Materialscientist, Mckenzie-brown-2310-2323, MelbourneStar, Melmann, Missvain, Mitch Ames, NeilN, Nemo bis, NickW557, Nymf, Oshwah, Perey, Philip Trueman, Pinethicket, Quenhitran, Rcehy, Retal, Scenariod, Senor Cuete, Ser Amantio di Nicolao, Serols, Shellwood, Smalljim, Smartsfordays, Steve03Mills, Synamicloser, Tassedethe, Tom.Reding, Trappist the monk, Turgan, Velella, Vieque, Vtc20165, Vtc4944, Wavelength, Wayne Roberson, Austin, Texas, Welcome to new york, Wiae, Widr, Zanaya lover234, Zeffi, Zppix, 174 anonymous edits 115

Aztec calendar *Source:* https://en.wikipedia.org/w/index.php?oldid=803675335 *License:* Creative Commons Attribution-Share Alike 3.0 *Contributors:* A.Skromnitsky, AgadaUrbanit, Agelix, Agyle, Alansohn, Atif.t2, Avoided, BD2412, Bender235, BethNaught, Bob the Wikipedian, Bollyjeff, Booblahbooblah, CJLL Wright, Calliopejen1, CanadianLinuxUser, Cavrdg, Chrism, ClueBot NG, Cnilep, CommonsDelinker, Copyeditor42, Courcelles, Danielmcg182, Deor, Disdero, Djembayz, ESkog, EmanWilm, Favonian, Fetchcomms, Flyer22 Reborn, Freyjali, Funandtrvl, Garrondo, GeneralizationsAreBad, Ginsuloft, Glacialfox, Goldenbrook, Grae Bear, HJPD, HamburgerRadio, Hamtechperson, Icairns, Icseaturtles, Ixfd64, Jarlz0r, Jim1138, K6ka, Kingpin13, Kwamikagami, KylieTastic, Lekro, Lowercase Sigma, MacedonianBoy, Madman2001, Makeemlighter, Martarius, Materialscientist, Maunus, Metatron 3.1, Michaelabril, Mild Bill Hiccup, Miniman879, Mistmaster45, Monkeyfox, MusikAnimal, Nekami, Nepenthes, Nihiltres, Orange Suede Sofa, Philip Trueman, Pinethicket, Polyxeros, Proofreader77, Ptbotgourou, Quinxorin, Racerx11, Reach Out to the Truth, Reaper Eternal, RenamedUser01302013, Rhododendrites, Rimarlag, Rosemania, Seaphoto, Second Quantization, Senor Cuete, Shirt58, Simon Burchell, Smyth, Some Gadget Geek, Stamptrader, Tassedethe, The High Fin Sperm Whale, The Thing That Should Not Be, Thincat, Tide rolls, Tisquesusa, Tom.Reding, Trappist the monk, Unitedelephant, Vanadium Zeppelin, Vrenator, Vvesikko, Wasell, Wavelength, Wayne Slam, Wwiipage, Xdnx, Yamaguchi先生, Zntrip, 248 anonymous edits 130

Astronomy in the medieval Islamic world *Source:* https://en.wikipedia.org/w/index.php?oldid=807009233 *License:* Creative Commons Attribution-Share Alike 3.0 *Contributors:* 468SM, AKS.9955, Akhooha, Al-Andalusi, Alangnixon, Alexqgb, AptitudeDesign, Aquib American Muslim, AstroLynx, BD2412, Bender235, Bgwhite, Bladesmulti, Bless sins, Bloodofox, Brandmeister, Chamboz, Chewings72, Chiswick Chap, Chris the speller, Chumbino, Citation bot 1, Citation bot 4, ClueBot NG, CommonsDelinker, Cyfal, David J Wilson, Dbachmann, Destroyer11623, Editor2020, Ego White Tray, Eteethan, Fences and windows, Flybywire e2c, Full Shunyata, Fæ, Galt42, Gattan, Glrx, GoingBatty, Gun Powder Ma, Hajatvrc, Headbomb, Hinio, Hmains, Hmainsbot1, Italia2006, J 1982, J8079s, Jagged 85, John of Reading, Jonathan.kade, Josve05a, KRY0F3N1X, Kaka Mughal, Kautilya3, Kgrad, Khazar2, Khirurg, LaurentianShield, LilHelpa, LindsayH, Machine Elf 1735, Magioladitis, Mandarax, Materialscientist, Maurice Carbonaro, Maxisi, MegA, Mgiganteus1, Michael Hardy, Mr Stephen, Ms Sarah Welch, Mughal Lohar, Muhammadahmad79, Muneebajmal1, Nannadeem, Nasa-verve, Nemoscis, Nfr-Maat, Ninney, Non-dropframe, OlEnglish, Onmywaybackhome, Orient Bengal, Philip Trueman, Piotrus, Ptbotgourou, Pushnell, Rarevogel, Rfassbind, Rich Farmbrough, Rjwilmsi, Rursus, Ruud Koot, SSR2000, Shonuff 1, Someone65, Spacepotato, SteveMcCluskey, Sunrise, Syncategoremata, TAnthony, The Anome, The Anomebot2, Thor Dockweiler, Tom.Reding, Tommy2010, TreasuryTag, Tyrol5, Urania's Muse, Wavelength, Webclient101, William M. Connolley, Wiqi55, Woohookitty, ZxxZxxZ, علي ويكي, ماني, 113 anonymous edits 139

Astronomia nova *Source:* https://en.wikipedia.org/w/index.php?oldid=803280259 *License:* Creative Commons Attribution-Share Alike 3.0 *Contributors:* (jarbarf), Andres, Atlantic306, Bambaiah, Bender235, Binksternet, Bobbie73, Bribett618, CAPTAIN RAJU, Caerwine, Chris the speller, Conscious, Cwkmail, Dantrom, Dolphin51, Druiffic, ELApro, Edybevk, Egsan Bacon, Euty, FisherQueen, Gaussbach, Good Olfactory, GregorB, Götz, Headbomb, Hugo999, Jaraalbe, Jerzy, John Nevard, Josve05a, Karlpoppery, Korovioff, Le Pied-bot~enwiki, Maestlin, MakeRocketGoNow, Marvin Diode, Mike s, Modest Genius, Mu301, Naraht, Nerd1a4i, Ohconfucius, Omnipaedista, Pegship, Polly Hedra, Ragesoss, Rjm at sleepers, STBotD, Salsb, Sandius, Stesmo, SteveMcCluskey, Tom.Reding, Trappist the monk, VoABot II, Wassermann~enwiki, Westwind2, Will Beback, Xanchester, Ywaz, 44 anonymous edits 157

Epitome Astronomiae Copernicanae *Source:* https://en.wikipedia.org/w/index.php?oldid=747465669 *License:* Creative Commons Attribution-Share Alike 3.0 *Contributors:* Bender235, Charles Matthews, Chiswick Chap, DDupard, Johnsoniensis, KiloByte, MusikAnimal, Omnipaedista, Valeria De Francesca (BEIC), 5 anonymous edits 163

Observational astronomy *Source:* https://en.wikipedia.org/w/index.php?oldid=793726591 *License:* Creative Commons Attribution-Share Alike 3.0 *Contributors:* Ageekgal, Akumiszcza, Altenmann, Anmclarke, Antihistoriaster, AstroLynx, BD2412, Bear-rings, Beland, Bergsten, BobQQ, Bobblewik, BotKung, Bota47, CDN99, ClueBot NG, Czj, D4g0thur, DaHuzyBru, Daniel Mietchen, David Adam Kess, David Eppstein, Dawnseeker2000, Deglr6328, Direwolf, Dr. Submillimeter, Déchiqueteuse, Elrond~enwiki, Emesee, Everyking, Flyer22 Reborn, G8ina, Geologist258, HMSSolent, Headbomb, Homonihilis, I am One of Many, J 1982, Jagged 85, Jakebarrington, Jmencisom, JohnOwens, K6ka, Karada, Karl-Henner, Karol Langner, Kozuch, Kungfuadam, KylieTastic, LDStNick, LL221W, Liamgilmartin, Lightmouse, Loraof, Lunokhod, Manishkrisna108, Markov, Melaen, Michael Hardy, Mike s, Mu301, Nasa-verve, Nesaverde, Norahsharon, Novangelis, Omnipaedista, Phe, RJHall, Randy Kryn, Rnt20, Roadrunner, Roberticus, Shanes, Shrutijm, Srikeit, Stuver, The Land, Thosjleep, TimBentley, Tom.Reding, Twinsday, Udorawala, Vanished user 5g6brbadLOiImSJylcnP, Vina-iwbot~enwiki, WilliamKF, Yurivict, 68 anonymous edits 167

Image Sources, Licenses and Contributors

The sources listed for each image provide more detailed licensing information including the copyright status, the copyright owner, and the license conditions.

Figure 1 *Source:* https://en.wikipedia.org/w/index.php?title=File:Newgrange_ireland_750px.jpg *License:* GNU Free Documentation License *Contributors:* Uploaded by Popsracer ... 2
Figure 2 *Source:* https://en.wikipedia.org/w/index.php?title=File:Auglish.jpg *License:* Creative Commons Attribution-Sharealike 2.0 *Contributors:* Alun Salt ... 3
Figure 3 *Source:* https://en.wikipedia.org/w/index.php?title=File:Uxmal01-panorama.jpg *License:* Creative Commons Attribution 2.0 *Contributors:* FlickrLickr, FlickreviewR, Glenn, MGA73bot2, RubeHM ... 4
Figure 4 *Source:* https://en.wikipedia.org/w/index.php?title=File:Chichen_Itza_Observatory_2_1.jpg *License:* GNU Free Documentation License *Contributors:* Fcb981 ... 9
Figure 5 *Source:* https://en.wikipedia.org/w/index.php?title=File:NAMA_Machine_d'Anticythère_1.jpg *License:* Creative Commons Attribution 2.5 *Contributors:* User:Marsyas ... 12
Figure 6 *Source:* https://en.wikipedia.org/w/index.php?title=File:Fajadadiagram.JPG *License:* Public Domain *Contributors:* Alan Liefting, Auntof6, Carlog3, OgreBot 2, Soerfm ... 13
Figure 7 *Source:* https://en.wikipedia.org/w/index.php?title=File:Decdiag600.jpg *License:* Creative Commons Attribution-ShareAlike 3.0 Unported *Contributors:* Alunsalt∼commonswiki, BotMultichill, Odysseus1479, Perhelion, 1 anonymous edits ... 15
Figure 8 *Source:* https://en.wikipedia.org/w/index.php?title=File:Earth_precession.jpg *License:* Public Domain *Contributors:* Robert Simmon, NASA GSFC ... 17
Figure 9 *Source:* https://en.wikipedia.org/w/index.php?title=File:Bayeux_Tapestry_scene32_Halley_comet.jpg *Contributors:* Myrabella ... 18
Figure 10 *Source:* https://en.wikipedia.org/w/index.php?title=File:Argo_Navis_Hevelius.jpg *License:* Public Domain *Contributors:* Johannes Hevelius ... 19
Figure 11 *Source:* https://en.wikipedia.org/w/index.php?title=File:1st_Pylon_Karnak_Temple.JPG *License:* Public domain *Contributors:* 1989, JMCC1, Neithsabes, Rémih, Tedmek, Wrightbus, 1 anonymous edits ... 21
Figure 12 *Source:* https://en.wikipedia.org/w/index.php?title=File:Newgrange_Eingang_Stein.jpg *License:* Creative Commons Attribution 2.5 *Contributors:* Clemensfranz ... 22
Figure 13 *Source:* https://en.wikipedia.org/w/index.php?title=File:All_Gizah_Pyramids.jpg *License:* Creative Commons Attribution-Sharealike 2.0 *Contributors:* Ricardo Liberato ... 23
Figure 14 *Source:* https://en.wikipedia.org/w/index.php?title=File:Senenmout-constellations.jpg *License:* Creative Commons Zero *Contributors:* bakha ... 24
Figure 15 *Source:* https://en.wikipedia.org/w/index.php?title=File:Serpent_head_at_the_base_of_El_Castillo.jpg *License:* Creative Commons Attribution 2.0 *Contributors:* Frank Kovalchek ... 25
Figure 16 *Source:* https://en.wikipedia.org/w/index.php?title=File:Summer_Solstice_Sunrise_over_Stonehenge_2005.jpg *License:* Creative Commons Attribution-Share Alike 2.0 Generic *Contributors:* 99of9, Acabashi, Ariadacapo, Carlos Luis M C da Cruz, Edward, J 1982, Julia W, Man vyi, Neukoln, RedWolf, Saperaud∼commonswiki, Solipsist∼commonswiki, Thesevenseas, Thierry Caro, Wereon, Winterkind, 4 anonymous edits ... 26
Figure 17 *Source:* https://en.wikipedia.org/w/index.php?title=File:PicMaesEntrance.jpg *License:* Creative Commons Attribution-Sharealike 2.0 *Contributors:* Islandhopper ... 27
Figure 18 *Source:* https://en.wikipedia.org/w/index.php?title=File:Uxmal-Palace-of-the-Governor.jpg *License:* Public Domain *Contributors:* Régis Lachaume ... 28
Figure 19 *Source:* https://en.wikipedia.org/w/index.php?title=File:Chaco_Canyon_Chetro_Ketl_great_kiva_plaza_NPS.jpg *License:* Public Domain *Contributors:* National Park Service (United States) ... 29
Figure 20 *Source:* https://en.wikipedia.org/w/index.php?title=File:Lascaux_01.jpg *License:* Creative Commons Attribution-ShareAlike 3.0 Unported *Contributors:* Peter80 ... 30
Image *Source:* https://en.wikipedia.org/w/index.php?title=File:Wiktionary-logo-en-v2.svg *Contributors:* User:Dan Polansky, User:Smurrayinchester 42
Image *Source:* https://en.wikipedia.org/w/index.php?title=File:Symbol_support_vote.svg *License:* Public Domain *Contributors:* Anomie, Fastily, Jo-Jo Eumerus ... 43
Figure 21 *Source:* https://en.wikipedia.org/w/index.php?title=File:Babylonian_tablet_recording_Halley's_comet.jpg *License:* Public Domain *Contributors:* 84user, Calliopejen1, File Upload Bot (Magnus Manske), Mmcannis∼commonswiki, OgreBot 2 ... 46
Image *Source:* https://en.wikipedia.org/w/index.php?title=File:Commons-logo.svg *License:* logo *Contributors:* Anomie, Callanecc, CambridgeBayWeather, Jo-Jo Eumerus, RHaworth ... 53
Image *Source:* https://en.wikipedia.org/w/index.php?title=File:Chaos_Monster_and_Sun_God.png *License:* Public Domain *Contributors:* Doug4, Ellin Beltz, Georgelazenby, Gerd Leibrock, JMCC1, KDS4444, Marcus Cyron, Pierre cb, Roberto.Amerighi, Vles1, 1 anonymous edits ... 54
Image *Source:* https://en.wikipedia.org/w/index.php?title=File:Astrological_Chart_--_New_Millennium.svg *Contributors:* User:Fred the Oyster 54
Image *Source:* https://en.wikipedia.org/w/index.php?title=File:Astrologia-tynkä.jpg *License:* Public Domain *Contributors:* Cathy Richards, Hautala, Hector Corrector, JMCC1, Theoteryi, Waldir, WolfgangRieger ... 55
Figure 22 *Source:* https://en.wikipedia.org/w/index.php?title=File:Babylon_relief.jpg *License:* Public Domain *Contributors:* Andre Engels, Bibi Saint-Pol, Erakis∼commonswiki, Jarekt, Jbribeiro1, Lusitana, Olivier, Rüdiger Wölk, Slowking4, Thiotrix, Wst, Zunkir ... 58
Figure 23 *Source:* https://en.wikipedia.org/w/index.php?title=File:Hindu_calendar_1871-72.jpg *License:* Public Domain *Contributors:* Abecedare, Cohesion, Durga, Ekabhishek, Mattes, Ranveig, Redtigerxyz, Roland zh ... 66
Figure 24 *Source:* https://en.wikipedia.org/w/index.php?title=File:Jantar_Mantar_at_Jaipur.jpg *License:* Public domain *Contributors:* Abhishekjoshi, BluesyPete, ClemRutter, JuergenG, Knowledge Seeker, Man vyi, Roland zh ... 70
Figure 25 *Source:* https://en.wikipedia.org/w/index.php?title=File:Jantar_Delhi.jpg *License:* Creative Commons Attribution-Sharealike 2.0 *Contributors:* Original uploader was Nikkul at en.wikipedia ... 70
Figure 26 *Source:* https://en.wikipedia.org/w/index.php?title=File:Grad-Scale-Hindu-Arabic-numerals.jpg *License:* Creative Commons Attribution 3.0 *Contributors:* soylentgreen23 on Flickr. ... 71
Figure 27 *Source:* https://en.wikipedia.org/w/index.php?title=File:Detail_from_jantar_mantar_jaipur.jpg *License:* Creative Commons Attribution-Sharealike 3.0 *Contributors:* ♣ ℓ u m i è r e ♣ on Flickr. ... 71
Figure 28 *Source:* https://en.wikipedia.org/w/index.php?title=File:AiKhanoumSunDial.jpg *License:* Creative Commons Attribution-ShareAlike 3.0 Unported *Contributors:* User:World Imaging ... 73
Figure 29 *Source:* https://en.wikipedia.org/w/index.php?title=File:NAMA_Machine_d'Anticythère_1.jpg *License:* Creative Commons Attribution 2.5 *Contributors:* User:Marsyas ... 79
Figure 30 *Source:* https://en.wikipedia.org/w/index.php?title=File:Anaximander.jpg *License:* Public Domain *Contributors:* Anne97432, Bibi Saint-Pol, JMCC1, Mattes, Sailko, SirSlither, Tomisti, Tony Rotondas, Un1c0s bot∼commonswiki, Warburg, 1 anonymous edits ... 80
Figure 31 *Source:* https://en.wikipedia.org/w/index.php?title=File:Ptolemaicsystem-small.png *License:* Public Domain *Contributors:* Fastfission 82
Figure 32 *Source:* https://en.wikipedia.org/w/index.php?title=File:Aristarchus_working.jpg *License:* Public Domain *Contributors:* Herbythyme, Jastrow, Konstable∼commonswiki, Leinad-Z∼commonswiki, Odysses∼commonswiki, Rocket000, Shakko, ZxxZxxZ, 2 anonymous edits ... 84
Figure 33 *Source:* https://en.wikipedia.org/w/index.php?title=File:AiKhanoumSunDial.jpg *License:* Creative Commons Attribution-ShareAlike 3.0 Unported *Contributors:* User:World Imaging ... 86
Figure 34 *Source:* https://en.wikipedia.org/w/index.php?title=File:Senenmut-Grab.JPG *License:* GNU General Public License *Contributors:* Neb-MaatRa ... 90
Figure 35 *Source:* https://en.wikipedia.org/w/index.php?title=File:Goddess_Nut_1.JPG *License:* Creative Commons Attribution-Sharealike 3.0 *Contributors:* Hans Bernhard (Schnobby) ... 90
Figure 36 *Source:* https://en.wikipedia.org/w/index.php?title=File:Nabta.gif *License:* Public Domain *Contributors:* BotMultichill, JMCC1, Xvdvoort, Zserghei ... 91
Figure 37 *Source:* https://en.wikipedia.org/w/index.php?title=File:Lepsius-228-III-ramsèsVI-détail.jpg *License:* Public Domain *Contributors:* Karl Richard Lepsius ... 92
Figure 38 *Source:* https://en.wikipedia.org/w/index.php?title=File:Astrological_Disc_LACMA_M.80.202.226_(1_of_3).jpg *Contributors:* Fæ, JMCC1 ... 93
Figure 39 *Source:* https://en.wikipedia.org/w/index.php?title=File:Dendera_Deckenrelief_08.JPG *License:* Creative Commons Attribution 3.0 *Contributors:* Olaf Tausch ... 94

Figure 40 *Source:* https://en.wikipedia.org/w/index.php?title=File:Dunhuang_star_map.jpg *License:* Public Domain *Contributors:* Brian J. Ford 97
Image *Source:* https://en.wikipedia.org/w/index.php?title=File:Chinese_rocket.png *License:* Public Domain *Contributors:* NASA 97
Figure 41 *Source:* https://en.wikipedia.org/w/index.php?title=File:Wide_View_of_the_Crab_Nebula.jpg *Contributors:* Jmencisom 99
Figure 42 *Source:* https://en.wikipedia.org/w/index.php?title=File:Su_Song_Star_Map_1.JPG *License:* Public Domain *Contributors:* PericlesofAthens 103
Figure 43 *Source:* https://en.wikipedia.org/w/index.php?title=File:YiXiangKaoCheng_02.JPG *License:* Public Domain *Contributors:* Mainly based on 徐光啓 's 《崇禎曆書》 (late Ming Dynasty) 106
Figure 44 *Source:* https://en.wikipedia.org/w/index.php?title=File:ChineseCelestialGlobe.JPG *License:* GNU Free Documentation License *Contributors:* Cold Season, Conscious, Foroa, Kilom691, MGA73bot2, Mike Peel, Nopira, Politikaner, Snowdog, Årvasbåo, 4 anonymous edits 108
Figure 45 *Source:* https://en.wikipedia.org/w/index.php?title=File:Observatoire_de_Peking.jpg *License:* Public Domain *Contributors:* Anville, Jean Baptiste Bourguignon d' 110
Figure 46 *Source:* https://en.wikipedia.org/w/index.php?title=File:Dengfeng_Observatory.jpg *License:* Creative Commons Attribution-Sharealike 2.0 *Contributors:* takwing.kwong 110
Image *Source:* https://en.wikipedia.org/w/index.php?title=File:Maya-Maske.jpg *License:* Creative Commons Attribution-Sharealike 3.0,2.5,2.0,1.0 *Contributors:* Wolfgang Sauber (User:Xenophon) 115
Image *Source:* https://en.wikipedia.org/w/index.php?title=File:MAYA-g-log-cal-D01-Imix.png *License:* Creative Commons Attribution 2.5 *Contributors:* User:CJLL Wright 118
Image *Source:* https://en.wikipedia.org/w/index.php?title=File:MAYA-g-log-cal-D11-Chuwen.png *License:* Creative Commons Attribution 2.5 *Contributors:* User:CJLL Wright 118
Image *Source:* https://en.wikipedia.org/w/index.php?title=File:MAYA-g-log-cal-D02-Ik.png *License:* Creative Commons Attribution 2.5 *Contributors:* User:CJLL Wright 118
Image *Source:* https://en.wikipedia.org/w/index.php?title=File:MAYA-g-log-cal-D12-Eb.png *License:* Creative Commons Attribution 2.5 *Contributors:* User:CJLL Wright 118
Image *Source:* https://en.wikipedia.org/w/index.php?title=File:MAYA-g-log-cal-D03-Akbal.png *License:* Creative Commons Attribution 2.5 *Contributors:* User:CJLL Wright 118
Image *Source:* https://en.wikipedia.org/w/index.php?title=File:MAYA-g-log-cal-D13-Ben.png *License:* Creative Commons Attribution 2.5 *Contributors:* User:CJLL Wright 118
Image *Source:* https://en.wikipedia.org/w/index.php?title=File:MAYA-g-log-cal-D04-Kan.png *License:* Creative Commons Attribution 2.5 *Contributors:* User:CJLL Wright 118
Image *Source:* https://en.wikipedia.org/w/index.php?title=File:MAYA-g-log-cal-D14-Ix.png *License:* Creative Commons Attribution 2.5 *Contributors:* User:CJLL Wright 118
Image *Source:* https://en.wikipedia.org/w/index.php?title=File:MAYA-g-log-cal-D05-Chikchan.png *License:* Creative Commons Attribution 2.5 *Contributors:* User:CJLL Wright 118
Image *Source:* https://en.wikipedia.org/w/index.php?title=File:MAYA-g-log-cal-D15-Men.png *License:* Creative Commons Attribution 2.5 *Contributors:* User:CJLL Wright 118
Image *Source:* https://en.wikipedia.org/w/index.php?title=File:MAYA-g-log-cal-D06-Kimi.png *License:* Creative Commons Attribution 2.5 *Contributors:* User:CJLL Wright 118
Image *Source:* https://en.wikipedia.org/w/index.php?title=File:MAYA-g-log-cal-D16-Kib.png *License:* Creative Commons Attribution 2.5 *Contributors:* User:CJLL Wright 118
Image *Source:* https://en.wikipedia.org/w/index.php?title=File:MAYA-g-log-cal-D07-Manik.png *License:* Creative Commons Attribution 2.5 *Contributors:* User:CJLL Wright 118
Image *Source:* https://en.wikipedia.org/w/index.php?title=File:MAYA-g-log-cal-D17-Kaban.png *License:* Creative Commons Attribution 2.5 *Contributors:* User:CJLL Wright 118
Image *Source:* https://en.wikipedia.org/w/index.php?title=File:MAYA-g-log-cal-D08-Lamat.png *License:* Creative Commons Attribution 2.5 *Contributors:* User:CJLL Wright 118
Image *Source:* https://en.wikipedia.org/w/index.php?title=File:MAYA-g-log-cal-D18-Etznab.png *License:* Creative Commons Attribution 2.5 *Contributors:* User:CJLL Wright 118
Image *Source:* https://en.wikipedia.org/w/index.php?title=File:MAYA-g-log-cal-D09-Muluk.png *License:* Creative Commons Attribution-ShareAlike 3.0 Unported *Contributors:* Bryan Derksen, CJLL Wright, Draceane 119
Image *Source:* https://en.wikipedia.org/w/index.php?title=File:MAYA-g-log-cal-D19-Kawak.png *License:* Creative Commons Attribution 2.5 *Contributors:* User:CJLL Wright 119
Image *Source:* https://en.wikipedia.org/w/index.php?title=File:MAYA-g-log-cal-D10-Ok.png *License:* Creative Commons Attribution-ShareAlike 3.0 Unported *Contributors:* BotMultichill, Bryan Derksen, CJLL Wright, ExtendedPartition, Ixtzib 119
Image *Source:* https://en.wikipedia.org/w/index.php?title=File:MAYA-g-log-cal-D20-Ajaw.png *License:* Creative Commons Attribution 2.5 *Contributors:* User:CJLL Wright 119
Image *Source:* https://en.wikipedia.org/w/index.php?title=File:Maya-Pop.jpg *License:* Public Domain *Contributors:* Japf 120
Image *Source:* https://en.wikipedia.org/w/index.php?title=File:Maya-Dresden-Yax.jpg *License:* Public Domain *Contributors:* Japf 120
Image *Source:* https://en.wikipedia.org/w/index.php?title=File:Maya-Dresden-wo.jpg *License:* Public Domain *Contributors:* Japf 120
Image *Source:* https://en.wikipedia.org/w/index.php?title=File:Maya-Dresden-Sak.jpg *License:* Public Domain *Contributors:* Japf 120
Image *Source:* https://en.wikipedia.org/w/index.php?title=File:Maya-Dresden-sip.jpg *License:* Public Domain *Contributors:* Japf 120
Image *Source:* https://en.wikipedia.org/w/index.php?title=File:Maya-Dresden-Keh.jpg *License:* Public Domain *Contributors:* Japf 120
Image *Source:* https://en.wikipedia.org/w/index.php?title=File:Sotz.jpg *License:* Public Domain *Contributors:* Japf 120
Image *Source:* https://en.wikipedia.org/w/index.php?title=File:Maya-Dresden-Mak.png *License:* Public Domain *Contributors:* Japf 120
Image *Source:* https://en.wikipedia.org/w/index.php?title=File:Maya-Dresden-Sek.jpg *License:* Public Domain *Contributors:* Japf 120
Image *Source:* https://en.wikipedia.org/w/index.php?title=File:Maya-Dresden-Kankin.png *License:* Public Domain *Contributors:* Japf 120
Image *Source:* https://en.wikipedia.org/w/index.php?title=File:Xul.png *License:* Public Domain *Contributors:* Japf 120
Image *Source:* https://en.wikipedia.org/w/index.php?title=File:Muan.jpg *License:* Public Domain *Contributors:* Japf 120
Image *Source:* https://en.wikipedia.org/w/index.php?title=File:Maya-Dresden-Yaxkin.jpg *License:* Public Domain *Contributors:* Japf 120
Image *Source:* https://en.wikipedia.org/w/index.php?title=File:Maya-Dresden-pax.jpg *License:* Public Domain *Contributors:* Japf 120
Image *Source:* https://en.wikipedia.org/w/index.php?title=File:Maya-Mol.png *License:* Public Domain *Contributors:* Japf 120
Image *Source:* https://en.wikipedia.org/w/index.php?title=File:Maya-Dresden-Kayab.png *License:* Public Domain *Contributors:* Japf 120
Image *Source:* https://en.wikipedia.org/w/index.php?title=File:Maya-Dresden-Chen.jpg *License:* Public Domain *Contributors:* Japf 120
Image *Source:* https://en.wikipedia.org/w/index.php?title=File:Maya-Dresden-kumku.jpg *License:* Public Domain *Contributors:* Japf 120
Image *Source:* https://en.wikipedia.org/w/index.php?title=File:Maya-Dresden-wayeb.jpg *License:* Public Domain *Contributors:* Japf 120
Figure 47 *Source:* https://en.wikipedia.org/w/index.php?title=File:East_side_of_stela_C,_Quirigua.PNG *License:* Public Domain *Contributors:* Googolplanck, Wagino 20100516, Павло Гетманцев, 2 anonymous edits 123
Image *Source:* https://en.wikipedia.org/w/index.php?title=File:Aztec_calendar.svg *License:* Creative Commons Attribution-Sharealike 3.0 *Contributors:* Unnamed Pre-Colombian Mexica artists of the 15th century. Keepscases modified it into . 130
Figure 48 *Source:* https://en.wikipedia.org/w/index.php?title=File:Monolito_de_la_Piedra_del_Sol.jpg *License:* Creative Commons Attribution-Sharealike 3.0,2.5,2.0,1.0 *Contributors:* El Comandante 131
Image *Source:* https://en.wikipedia.org/w/index.php?title=File:Cipactli.jpg *License:* Public Domain *Contributors:* Kilom691, Ptcamn∼commonswiki 132
Image *Source:* https://en.wikipedia.org/w/index.php?title=File:Ehecatl2.jpg *License:* Public Domain *Contributors:* Gryffindor, Kilom691, Ptcamn∼commonswiki 132
Image *Source:* https://en.wikipedia.org/w/index.php?title=File:Calli.jpg *License:* Public Domain *Contributors:* Kilom691, Ptcamn∼commonswiki 132
Image *Source:* https://en.wikipedia.org/w/index.php?title=File:Cuetzpalin.jpg *License:* Public Domain *Contributors:* Gospodar svemira, Kilom691, Ptcamn∼commonswiki 132
Image *Source:* https://en.wikipedia.org/w/index.php?title=File:Coatl.jpg *License:* Public Domain *Contributors:* Gospodar svemira, Kilom691, Ptcamn∼commonswiki, Tortillovsky 132
Image *Source:* https://en.wikipedia.org/w/index.php?title=File:Miquiztli.jpg *License:* Public Domain *Contributors:* Gospodar svemira, Kilom691, Ptcamn∼commonswiki 132
Image *Source:* https://en.wikipedia.org/w/index.php?title=File:Mazatl.jpg *License:* Public Domain *Contributors:* Kilom691, Ptcamn∼commonswiki 132
Image *Source:* https://en.wikipedia.org/w/index.php?title=File:Tochtli.jpg *License:* Public Domain *Contributors:* Kilom691, Ptcamn∼commonswiki 132
Image *Source:* https://en.wikipedia.org/w/index.php?title=File:Atl3.jpg *License:* Public Domain *Contributors:* CommonsDelinkerHelper, Kilom691, Ptcamn∼commonswiki 132

License

Index